Fish Genetics and Aquaculture Biotechnology

Fish Genetics and Aquaculture Biotechnology

Editors

T.J. Pandian
C.A. Strüssmann
M.P. Marian

SCIENCE PUBLISHERS, Inc.
Post Office Box 699
Enfield, New Hampshire 03748
United States of America

Internet site: *http://www.scipub.net*

sales@scipub.net (marketing department)
editor@scipub.net (editorial department)
info@scipub.net (for all other enquiries)

Library of Congress Cataloging-in-Publication Data

Fish genetics and acquaculture biotechnology/editors, T.J. Pandian,
C.A. Strüssmann, M.P. Marian.
p. cm.
Some papers were presented at the International Conference on
"Advanced Technologies in Fisheries and Marine Sciences".
Includes bibliographical references and index.
ISBN 1-57808-372-9
1. Fishes–Breeding–Congresses. 2. Fishes–Genetic Engineering–Congresses. 3. Aquaculture–Congresses, I. Pandian, T.J. II. Strüssmann, C.A. III. Marian, M.P.

SH 155.5.F57 2004
639.3-dc22

2004056498

ISBN 1-57808-372-9

Published by Science Publishers, Inc., Enfield, NH, USA
Printed in India.

Preface

Fish Genetics and Aquaculture Biotechnology is an outcome of the International Conference on "Advanced Technologies in Fisheries and Marine Sciences" organized by the Center for Marine Sciences and Technology, Manonmaniam Sundaranar University. Representative contributions from the sessions on Transgenesis, Chromosome Engineering and Sex Control, Molecular Endocrinology, Microbal Diseases and Vaccines, and Bioactive Compounds have been selected in order to bring about a comprehensive volume. Of the 12 selected articles, six are original contributions and the rest are comprehensive reviews. Hence, the book has a proporationate blend of original as also review contributions.

The problem of feeding the world's 7-8 billion inhabitants through agriculture in an environmentally sustainable manner has prompted scientists to look upon water as the major source of food production. Although selective breeding is an excellent method for strain improvement of crops and farm animals so as to increase productivity, it requires a relatively long period of time. Moreover, the mixing of the entire genome may transmit unwanted traits to the progeny. Besides evolutionary differences might pose a barrier to the transmission of advantageous genes between species. Transgenesis, however, allows the introduction of novel genes or augmentation of a particular character by merely transferring the manipulated copies of the gene from the same or different species. For successful generation of useful transgenics, indigenous vectors and effective methods of transfer are essential. Presenting the status of our knowledge in fish transgenics, the first chapter describes the production of indigenous bi-cistronic vector and transgenic rohu growing 4-5 times faster than the controls. Microinjection is a time-consuming, laborious, species-specific method of gene transfer while other methods of gene transfer such as electroporation also possess certain disadvantages. Using Nuclear Signal Localization Peptides non-covalently complexed to plasmid DNA, Alestrom and his colleagues have achieved germline integration and expression of single copies of the luciferase reporter gene in zebrafish.

Fishes are known for their amazing ability to tolerate various levels of ploidy, from haploid to heptaploid, unequal and even uniparental genomic contribution to the zygote as well as genetic contributions from parents of different species. Pandian and Kirankumar have comprehensively reviewed the relevant literature on androgenesis in fish. They have restored the genome of rosy barb using its cadaveric sperm and genome-inactivated eggs of tiger barb. In a thought-provoking contribution, Strüssmann and his colleagues have described the currently available methods of sex control and novel hypotheses on the mechanisms of gonadal sex differentiation in fishes. Production of monosex progenies is expected to contribute to increased productivity and profitability in fish farming and also minimize the environmental impact of the introduction of exotic species.

The sustainability of commercial aquaculture depends on the continuous and predictable supply of seeds of cultivated fish species. In capitivity, a majority of farmed fishes fail to reproduce successfully due to the failure in releasing lutenizing hormone. In a most thorough and excellent review, Alok and Zohar have elucidated the endogenous GnRH-GnRH-R interaction that blocks the gonadotropin release in fishes bred under captivity. In shrimp culture too, it has become increasingly difficult to procure an adequate number of quality seeds from wild-caught brooders. The most common method of induced spawning in shrimp is the unilateral ablation of the eyestalk, with the removal of the source of hormones that inhibit gonadal development. Typically, 70% of wild-caught brooders can be induced to spawn by this technique, while in domesticated brooders, this value is less than 20%. In a refreshingly new approach, Wilson and her team have replaced the eyestalk ablation with passive immunization, i.e. the use of an antibody that specifically removed the gonad-inhibiting hormone from circulation while leaving the other hormones unaffected. Following this viewpoint, Vishakan and his colleagues have demonstrated the possibility of homeopathic induction of spawning in goldfish.

Fishes are very vulnerable to microbial diseases. In spite of many technological advances, it is still very difficult to accurately diagnose pathogens. For example, there are over 200 serotypes of *Vibrio cholerae,* of which, only serotypes 01 and 139 cause cholera. Occurrence of vibriosis in penaeid shrimp is generally influenced by factors such as stress, environmental contamination and the abundance of potential pathogenic bacteria. Abdel-Aziz and Das have made an attempt to describe the antibiotic sensitivity of *Vibrio harveyi* biotypes isolated from pond-reared *Penaeus indicus* collected in Saudi Arabia. Also, in a fairly thorough review, John and his colleagues have presented an account of our knowledge on DNA vaccines, their advantages, the methods of administration and safety issues.

Marine organisms are regarded as an unlimited source of new bioactive compounds. Gerdts and his colleagues have focused on an exciting facet of the interaction between a sponge, *Halichondria panicea* and its accompanying bacteria. Thus, they have described the existing methods of isolation and cultivation of microbes by simulating natural conditions.

We sincerely hope that this book may serve as a source of information and stimulation for further work. We record our grateful appreciation to the authorities of our universities for their support.

Spring, 2004

T.J. Pandian
C.A. Strüssmann
M.P. Marian

Contents

Preface *v*

1. Contribution to Transgenesis in Indian major carp *Labeo rohita* 1
T.J. Pandian and *T. Venugopal*

2. Gene Transfer to Germline and Somatic Tissues of Zebrafish 21
P. Aleström, P. Collas, J. Torgersen, M-R. Liang, A. Arenal and *R.N. Lillabadi*

3. Application of RAPD and AFLP to Detect Genetic Variation in Fish 29
P. Jayasankar

4. Androgenesis and Conservation of Fishes 37
T.J. Pandian and *S. Kirankumar*

5. Methods of Sex Control in Fishes and an Overview of Novel Hypothesis Concerning the Mechanisms of Sex Differentiation 65
C.A. Strüssmann, M. Karube, L.A. Miranda, R. Patiño, G.M. Somoza, D. Uchida, and *M. Yamashita*

6. Genes for Fish GnRHs and Their Receptor: Relevance to Aquaculture Biotechnology 81
D. Alok and *Y. Zohar*

7. Biotechnology to Improve Reproductive Performance and Larval Rearing in Prawns and Rock Lobsters 103
K. Wilson, M. Hall, M. Davey, M. Kenway and *D. Coren*

8. Homeopathic Induction of Spawning in Ornamental Fish 119
R. Vishakan, S. Balamurugan, C. Maruthanayaga and *P. Subramanian*

9. Asymptomatic Mortality of *Penaeus indicus* Infected by Non-luminescent *Vibrio harveyi* 123
E.S. Abdel-Aziz and *S.M. Dass*

10. DNA Vaccine Technology: Current Status and Potential Application in Aquaculture Industry 131
J.A. Chrisotopher John, S. Murali, M. Peter Marian and *Chi-Yao Chang*

11. Isolation of Antibody-like Substances from Marine Algae 145
U. Barros, and *A. Himanshu*

12. Intracellular and Associated Marine Bacteria in the Sponge *Halichondria panicea*: A Potential for Pharmaceuticals 151
G. Gerdts, A. Wichels, H. Doepke and *C. Schuett*

Index 161

Chapter 1

Contribution to Transgenesis in Indian major carp *Labeo rohita*

T.J. Pandian[1] and **T. Venugopal**[2]
[1]School of Biological Sciences, Madurai Kamaraj University, Madurai 625 021, India Email: tjpandian@eth.net
[2]Department of Medicine, Medical School, University of Massachusetts Worcester, MA 01605-2324, USA Email: venu3333@yahoo.com

ABSTRACT

Fishes are the more advantageous models for transgenesis than mammals. Of the 8,500 known genes and DNA sequences of piscine origin, only 101—mostly growth hormone genes—have been inserted into vectors. The significance of promoters and reporters in these vectors has been described. Internal ribosomal entry sites (IRES) element, known to promote internal initiation of RNA translation and to facilitate the expression of two or more proteins from a polycistronic transcript in eukaryotic cells, has been used for the first time in the construction of vectors containing growth hormone gene of Indian fishes such as rohu and catfish. The fragile and swelling eggs of rohu necessitated standardization of the sperm-mediation gene transfer technique to administer the vector pgcβ-rGH-IRES-EGFP into the eggs of rohu. Southern analysis confirmed genomic integration in 15% of the tested rohu belonging to the family lines 2 and 3 and another 25% of the juveniles were also proved transgenic but with the transgene persisting extra-chromosomally for longer than 1 year, perhaps due to the presence of replicon in the vector. Transgenics grew 4-5 times faster than the control.

Key words: Indigneous vectors, IRES element, sperm-mediated transfer, genomic integration, five times faster growth.

INTRODUCTION

Although selective breeding has proven to be an excellent method for strain improvement of crops and farm animals, it suffers from certain disadvantages. This strategy requires a relatively longer time, and the entire genome mixing may transmit unwanted traits. Also, selective breeding does not allow transmission of genes/traits from another species (Pandian, 1995). Transgenesis, however, allows the introduction of a novel gene or augmentation of a particular character by just transferring the manipulated copies of the gene from the same or different species. The first transgenic animal 'supermouse' was produced by direct injection of the growth hormone gene into the pronucleus of murine eggs (Palmiter et al., 1982). This scientific breakthrough triggered a large number of attempts to transfer gene in animals like nematode (Stinchcomb et al., 1985), fruit fly (Rublin and Spradling 1982), sea urchin (MaMahan et al., 1984), frog (Etkin et al., 1984) and farm animals such as rabbit, sheep, cattle and pig (Hammer et al., 1985; Church, 1987; Clark et al., 1987). The very first gene transfer attempt in fish was made by Zhu et al., (1985) in goldfish. Since then, many researchers have attempted gene transfer in fish and some of them have produced transgenics with accelerated growth (Du et al., 1992; Devlin et al., 1994; Rahman et al., 1998; Sheela et al., 1999; Nam et al., 2001).

GENETIC IMPROVEMENT OF *LABEO ROHITA*

Rohu exhibits a relatively slow growth compared to the other Indian major carps such as mrigal and catla, probably due to inbreeding (Reddy et al., 2001). In order to accelerate the growth rate, a selective breeding program was undertaken in India at the Central Institute of Freshwater Aquaculture (CIFA), Bhubaneswar, in collaboration with the Institute of Aquaculture Research (AKVAFORSK), Norway. Five wild stocks were collected from different Indian river systems, namely Ganga, Gomathi, Yamuna, Sutlej and Brahmaputra, along with the CIFA farmed wild stock. All these five stocks were crossed with each other and the better-growing hybrids were selected by following the family selection method, and the offsprings were studied for their growth performance.

After successive selective breeding attempts, the selectively bred line was named as 'jayanti rohu' and was evaluated in different agro-climatic conditions of the country. When the growth was compared with the locally farmed variety of the respective region, the 'jayanthi rohu' exhibited 22% selection response per generation (Reddy et al., 2001). Ways to distribute the genetically superior rohu seed to the fish farmers are being worked out.

FISH AS A MODEL SYSTEM

Fishes have been considered to be the best vertebrate model system for several advantages over the mammalian model (Pandian and Marian, 1994). This is primarily because: (i) fishes produce a larger number of gametes than mammals; (ii) fishes entail easier handling of gametes and non-requirement of re-implantation associated with external fertilization; and (iii) fish eggs are relatively more totipotent than that of mammals. Additionally, several highly homologous strains are available, along with several methods for obtaining new strains. A large number of mutants have also been generated. Unlike other vertebrates, gynogenetically reproducing species are also available in fishes (Pandian and Koteeswaran 1998).

Zebrafish has been established as one of the best vertebrate model systems for systematic mutageneic studies. More than 2000 mutations affecting the normal development have been identified (Postlethwait and Talbot, 1997).

Table 1.1: Key achievements and limitations in fish transgenesis

Authors	*Achievements*	*Limitations*
Zhu et al. (1985)	First to initiate and claim production of transgenic fish	No molecular evidence for integration, expression and transmission
Ozato et al. (1986)	First to microinject into oocyte nucleus and prove expression immunohistochemically	Reported mosaic distribution of exogene. Transmission studies not undertaken
Stuart et al. (1988)	First to undertake detailed germ-line transmission studies	Low frequency of transmission and the transgenics suffer from mosaicism
Zhang et al. (1990)	First to use fish growth hormone gene sequence for gene transfer, expression and inheritance	Mosaic individuals produced; mRNA levels not studied
Inoue et al. (1990)	First to report successful gene transfer by electroporation	Low integration frequency; expression not studied
Tamiya et al.(1990)	First to use marker gene luciferase in transgenic fish studies	Transient expression; no evidence for integration and transmission
Khoo et al. (1992)	First attempt to use sperm cells as vector to introduce genes into fish egg	Reported only extrachromo-somal persistence but no expression
Du et al. (1992)	First report on rapidly growing transgenic fish using "all-fish" gene construct	Maturation of fish and inheritance of enhanced growth not established
Muller et al. (1992)	First to carry out sperm electro-poration for gene transfer	No demonstration of long-term persistence or integration
Devlin et al. (1994)	First to show extraordinary growth using "all-salmon" construct	F_1 suffered mosaicism
Sheela et al. (1998)	First to confirm the genomic integration, by successive fin clipping for 7 times	Correlation was not found between reporter gene and growth hormone gene
Nam et al. (2001)	First to generate a transgenic fish displaying dramatic growth acceleration and exceptionally larger size	
Venugopal et al. (2004)	First to show IRES elements could be used for bi-cistronic vector construction of GH cDNA for research in transgenic fish	

The available vertebrate models have larger genomes, measuring thousands of mega bases. However, *Fugu rubripes*, commonly known as pufferfish, has 7.5 times smaller genome than that of human, which consists of 400 mb (0.4 - 0.5 pg) DNA. Hence, *Fugu* is regarded as the simplest vertebrate model system in terms of genome size (Brenner et al., 1993; Elgar et al., 1996). The entire genome of zebrafish and *Fugu* is currently being sequenced.

GROWTH ENHANCEMENT OF TRANSGENIC FISH

In 1921, Evans and Long (1921) were the first to describe the growth-promoting function of the pituitary gland. The human Growth Hormone (GH) encoding cDNA was perhaps the first to be isolated and characterized (Li and Evans 1944). Growth hormone, Chorionic Somatomotropin (CS), Placental Lactogen [PL]) and Prolactin (PRL) are all a family of hormones, thought to have evolved from a common precursor (Nial et al., 1971; Miller and Eberhardt, 1983). In vertebrates, growth hormone controls the post-natal somatic growth (Sakamoto et al., 1993) and reproduction (Van der Kraak et al., 1990; Le Gac et al., 1993). These functions of GH are mediated by Insulin-like Growth Factors (IGF 1 and 2) (Mathews et al., 1986; Sara, 1991) and growth-related protein kinases (Ralph et al., 1990).

GH administration has shown to accelerate the growth rate in a number of animals. Numerous scientists (Du et al., 1992; Devlin et al., 1994; Rahman et al., 1998; Sheela et al., 1999; Nam et al., 2001) have attempted to generate fast-growing transgenic food-fish. Hence, cloning, characterization and expression of GH have been the subject of extensive research (e.g. Chang et al., 1992) during the last decade. Due to its importance in fish culture, GH encoding cDNA has so far been cloned from about 30 fish species (Bernardi et al., 1993; Pandian and Marian, 1994; Pandian et al., 1999).

GROWTH-ENHANCED TRANSGENIC FISH

Intensive research on transgenesis during the last 15 years has transformed the dream of producing growth-accelerated transgenic fish into a reality. Table 1.1 lists the landmark events in transgenic fish research. Recently, transgenic salmon, namely AquaAdvantageTM, which grows 4-6 times faster but consumes 20% less food than the control, has been generated in Canada and the USA, and is being licensed to farmers for breeding (www.aquabountyfarms.com). Attempts are in progress to apply the same technology to other finfishes like the Arctic charr, trout, tilapia, turbot and halibut. Fish is very likely to be the first transgenic animal to be commercialized.

Indigneous vectors

In most gene transfer studies on fishes, the transferred gene was the gene of availability rather than that of choice. A variety of genes originating from bacteria to mammal have been used for want of suitable genes of piscine origin (Pandian et al., 1994). But achieving a good expression of heterologous gene/introns has remained a problem. It was thought that due to inappropriate splicing and translation, the heterologous (e.g. mammalian) introns may not be efficiently processed by the piscine cellular machinery (Friedenreich and Schartl, 1990). The ever increasing knowledge on structure and function of eukaryotic gene has shown the need for introns, enhancer, boundary and locus control regions (Iyengar et al., 1996), besides a suitable promoter in the construction of an appropriate vector for use in transgenic fish system. Thus, considerable work undertaken during the last 15 years has led to the cloning and characterization of genes of piscine origin. Due to current advancements in molecular biology, more than 8,500 genes and cDNA sequences of piscine origin have so far been isolated (Biswas et al., 2001), and the number is increasing by the day. However, only about 101 out of 5,116 constitute commercially important genes such as somatotropin hormone family (Table 1.2), and less than one fourth of them have been inserted into vectors and can be treated in gene transfer studies. A glance over the history of these commercially important genes reveals that the trend of using heterologous sequences in the mid 1980s began to lean towards homologous sequences in the early nineties. Since 1995, the use of 'all-fish' constructs has become a practice.

Table 1.2 : Most represented protein coding sequences of teleost piscine origin (Biswas et al., 2001)

Name of the DNA sequence	*Number*	*Reference**
Cytochrome b/b6	1794	IPR000179
Class II histocompatibility antigen, β-chain, β-1 domain	374	IPR000353
Immunoglobulin and major histocompatibility complex domain	316	IPR003006
NADH-Ubiquinone/Plastoquinone (complex I), various chains	306	IPR001750
Homeobox domain	303	IPR001356
Rhodopsin-like GPCR superfamily	293	IPR000276
Major histocompatibility complex protein, Class I	292	IPR001039
Cytochrome C oxidase subunit I	229	IPR000883
Eukaryotic protein kinase	166	IPR000719
Tyrosine kinase catalytic domain	129	IPR001245
Opsin	126	IPR001760
ATP synthase A subunit	109	IPR000568
Globin	109	IPR000971
Somatotropin hormone family	101	IPR001400
C-4 type steroid receptor zinc finger	94	IPR001628
Lambda and other repressor helix-turn-helix	88	IPR000047
Rhodopsin	78	IPR000732
Helix-loop-helix dimerization domain	71	IPR001092
Cytochrome P450 enzyme	70	IPR001128
EF-hand family	68	IPR002048

* http://www.ebi.ac.uk/interpro/

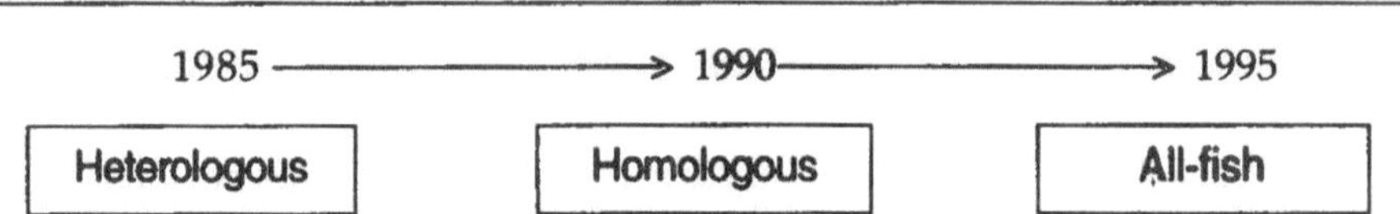

Table 1.3 : List of piscine genes/promoters available in the form of vectors of commercial fish

Promoter	*Species*	*Gene construct*	*Reference*
β-actin promoter	Common carp	pCaβ-CAT	Depoint-Zilli et al. (1988)
	Common carp	pCaβ-CAT	Hew (1989)
	Common carp	pCaβ-csGHcDNA	Liu et al. (1990)
	Common carp FV1 and FV2	pCaβ-csGH	MaClean et al. (1992)
	Common carp	pCaβ-rtGH	Chen et al. (1993)
	Common carp	pCaβ-CAT	Hackett et al. (1994)
	Common carp	pCaβ-csGHcDNA	Moav et al. (1995)
	Common carp	pCaβ-LacZ	Alam et al. (1996)
	Mudloach	pLβa-LGH	Nam et al. (2001)
Metallothionein promoter	Rainbow trout	prtMta-CAT-β-gal	Inoue et al. (1990)
	Rainbow trout	prtMta-CAT prtMta-CAT	Winkler et al. (1992)
	Rainbow trout	prtMta-CAT prtMta-CAT	Hong et al. (1993)
	Rainbow trout	prtMtb-gbs-GHcDNA prtMta-gbs-GHcDNA	Cavari et al. (1993)
	Sockeye salmon	MTb gene	Chan and Devlin (1993)
	Sockeye salmon	POnMT-csGH	Devlin et al. (1994)
Antifreeze protein gene-promoter-region	Winter flounder	pPAFP-asGH	Fletcher et al. (1988)
	Winter flounder	pPAFP-AFP	Fletcher et al. (1988)
	Atlantic salmon	pasAFP-AFP	Hew et al. (1991)
	Winter flounder	pPAFP-CAT	Hew et al. (1991)
	Atlantic salmon	pPAFP-GH	Du et al. (1992)
	Winter flounder	pOpAFP-GHcDNA	Tseng et al.(1994)
	Ocean pout	pOpAFP-GHcDNA	Gong et al. (1994)
	Ocean pout	pGnRH-LUC	Erdelyi et al. (1994)
Gonadotrophin-releasing hormone region	Salmon	pα- globin gene	Alestrom et al. (1991)
α-globin gene	Common carp	prtH3-CAT	Yoshizaki et al. (1991)
Histone promoter	Rainbow trout	H3 gene	Muller et al. (1992)
Histone H3	Sockeye salmon	Protamine gene	Chan and Devlin (1993)
Protamine	Sockeye salmon	ZpβgpGH	Chan and Devlin (1993)
Zona pellucida	Winter flounder	ZpβrtGH	Sheela et al. (1998)
Zona pellucida	Winter flounder		Sheela et al. (1999)

Table 1.3 shows the list of genes available in the form of vectors. A careful survey of these vectors clearly states that the available constructs are mostly for the economically important fish belonging to developed countries. It has become a necessity for the developing countries—which depend on fisheries

as an important food source—to develop 'all-fish' gene constructs for growth enhancement in fish species that are economically important to them (Pandian et al., 1999). Among the developing countries, China (e.g. *Cyprinus carpio*; Chiou et al., 1990) and Israel (*Sparus aurata*; Funkenstein et al., 1991) have attempted to construct vectors for their indigenous fish species.

In India, the need for production of 'swadeshi' gene constructs has been adequately emphasized (Pandian et al., 1994). Thus, the prime aim of the Indian research group was to clone and characterize the growth hormone (GH) encoding cDNA of rohu, a commercially important fish of India and to construct indigneous vectors to be used in transgenic fish research. The GH cDNA of Indian major carps *Labeo rohita*, *Cirrhina mrigala* and *Catla catla* (Venugopal et al., 2002) and Indian catfish *Heteropneustes fossilis* (Anathy et al., 2001) have been cloned and characterized. Recently, the homologous vectors have been transferred into eggs of respective species by suitable methods. A fast-growing autotransgenic rohu (Venugopal et al., 2004) has been generated as a result of these endeavors.

Reporter genes

Many reporter genes have been used in fish for better understanding of transgene integration (Khoo et al., 1992; Nam et al., 1999), expression (Rahman et al., 2000) and transmission (Khoo et al., 1992; Nam et al., 1999). The methods used to detect and quantify these reporter genes are relatively simple, rapid, less cumbersome than required for other genes such as growth hormone. Phenotypic growth has also been quantified as a measure of expression of the growth hormone transgene in the transgenic animal (Pandian et al., 1991; Du et al., 1992; Devlin et al., 1994; Sheela et al., 1998). However, growth is a complicated trait influenced by several factors other than the level of ectopically expressed transgene growth hormone and cannot always be taken as a measure of the transgene expression (Sheela et al., 1999). For instance, even the plasma level of growth hormone cannot serve as a reliable quantitative index of growth, as there is no correlation between the hormonal level and growth of the transgenic Atlantic salmon (Du et al., 1992). To simplify the identification procedures of growth hormone transgene expression, few research groups have constructed gene sequences containing a growth hormone gene and a reporter gene along with other usual elements (Sheela et al., 1998). Therefore, it is considered that a combination of growth hormone and reporter genes may minimize the cost and time required to identify putative founder transgenic fish. However, it is likely that a reporter gene like β-gal may not serve as quantitative index for the integration and expression of a fused transgene. For instance, Sheela et al. (1999) electroporated a gene construct containing yellow porgy GHcDNA with β-gal reporter gene into the eggs of *Heteropneustes fossilis*. The level of expression of GH gene, as quantified by the somatic growth of the putative transgenic fish, was 30% to 60% higher than control, whereas on using levels of reporter enzymes as a quantitative index, it was found to be 200% to 600% higher. Clearly, the expression of GH gene and the reporter gene was not proportionately equal.

Bi-cistronic vectors

Recently, a new vector system has been developed to express two different genes under a single promoter. Bi-cistronic vectors, in which the first gene is translated in a 5'cap-independent manner from mRNA, are employed with a variety of expression systems from cultured cells to transgenic animals (Mountford and Smith, 1995). Co-expression of two genes, where one is a marker/cell-surface antigen, and the second is the gene of interest, are needed in genetic engineering. The ability of the Internal Ribosome Entry site (IRES) element to promote internal initiation of RNA translation (Hellen and Sarnow, 2001) has facilitated the expression of two or more proteins from a polycistronic transcript in eukaryotic cells (Martinez-Salas, 1999; Fahrenkrug et al., 1999).

In cloning and construction of vectors of piscine origin, the IRES element has, however, so far not been used for transgenesis, although Wang et al. (2000) have used the IRES-EGFP construct to trace the exogenously introduced mRNA by tracing the EGFP expression in zebrafish embryos. In the present study, the IRES element has been used to generate a transgenic fish for the first time. The GH cDNA of rohu has been cloned in the vector and its fidelity has also been checked in zebrafish embryos (Venugopal et al., 2002). By demonstrating the effective use of IRES element for bi-cistronic expression in putative transgenic fish, this study has opened a new trend in transgenic biology.

METHOD OF GENE TRANSFER

Despite the development of different new methods, such as the use of retroviral integration protein (Ivics et al., 1993), nuclear localization signals (Collas and Alestrom, 1998; Liang et al., 2000) and particle gun bombardment (Yamauchi et al., 2000), the widely used methods for gene transfer in fish are microinjection (Ozato et al., 1986; Pandian et al., 1991) and electroporation (Inoue et al., 1990; Sheela et al., 1998, 1999). Microinjection is time-consuming, laborious, species-specific and, in some cases, technically demanding (Pandian and Marian, 1994). But even then, it is still widely used and ensures genomic/extrachromosomal persistence of transgene in 20% to 90% fish of different ages ranging from 3 weeks to 2 years (Table 1.4).

The next best method for transferring the transgene into fish eggs is electroporation. Although the over all frequency of DNA integration in transgenic fish produced by electroporation was equal to or slightly higher than that of microinjection, the actual amount of time required for handling a large number of embryos by electroporation is orders of magnitude less than the time required for the former proceses (Chen et al., 1996). Powers et al. (1992) have demonstrated that electroporation was superior to microinjection in

Table 1.4: Methods for gene transfer and gene transfer efficiency in fish

Method	*Species*	*Persistence of foreign DNA*		*Reference*
		Period	*(%)*	
1. Microinjection (i) Cytoplasm direct	Channel catfish	3 week	*20	Dunham et al. (1987)
	Medaka	1 month	90	Chong and Vielkind (1989)
	Carp	3 month	*6	Zhang et al. (1990)
	Channel catfish	Fry	2	Hayat et al. (1991)
	Northern Pike	2 month	6	Gross et al. (1991)
	Nigrobuna	2 years	20	Ueno et al. (1994)
	Indian catfish	Juvenile	24	Marian et al. (1997)
	Zebrafish	F_1 and F_2		Sheela et al. (1993)
	Zebrafish	3 month	35	Powers et al. (1992)
	Zebrafish	Fry	#69	Pandian et al. (1991)
	Common carp	3 month	0 – 16	Powers et al. (1992)
	Goldfish	Fry	*50	Zhu et al. (1985)
	Channel catfish	3 month	0 – 33	Powers et al. (1992)
(ii) Cytoplasm via micropyle	Tilapia	3 month	*67	Brem et al. (1988)
	Zebra cichlid	6-12 month	14	Marian (1995)
	Atlantic Salmon	8-11 month	*2-3	Du et al. (1992)
(iii) Cytoplasm 2 steps	Tilapia	3 month	18	Brem et al. (1988)
	Rainbow trout	6-12 month	*75	Chourrout et al. (1986)
	Atlantic salmon	8-11 month	4	Fletcher et al. (1988)
	Atlantic salmon	1 year	56	Rokkones et al. (1989)
	Rainbow trout	1 year	*39	Yoshizaki et al. (1991)
	Rainbow trout	70 day	*60	Yoshizaki et al. (1991)
(v) Nucleus	Medaka	2 day	50	Ozato et al. (1986)
2. Electroporation of fertilized eggs	Medaka	1 day	62	Tsai et al. (1995)
	Medaka	20 day	4	Inoue et al. (1990)
	Zebrafish	1 week	*65	Buno and Linser (1992)
	Loach	1 month	*62.5	Xie et al. (1993a,b)
	Common carp	3 month	*15-75	Powers et al. (1992)
	Channel catfish	3 month	0-100	Powers et al. (1992)
	Zebrafish	3 month	*15-75	Powers et al. (1992)
	Zebrafish	Fry	27	Zhao et al. (1993)
	Indian Catfish	Juvenile	*27	Marian et al. (1997)
	Zebrafish	4 week	*53	Sheela et al. (1998)
	Catfish	1 year	*15-100	Sheela et al. (1999)
3. Sperm as carrier (i) with electroporation	Carp	2 week	2.6	Muller et al. (1992)
	Labeo rohita	2 week	25	Venugopal et al. (1998b)
	Cirrhina mrigala	2 week	23	Venugopal et al. (1998b)
	Catla catla	2 week	13	Venugopal et al. (1998b)
	Tilapia	1 month	3	Muller et al. (1992)
	African catfish	10 day	3.5	Muller et al. (1992)
	Chinook salmon	Fry	1.5	Sin et al. (1993)
	Loach	2 week	*37.5	Tsai et al. (1995)
	Zebrafish	2 week	14.5	Patil and Khoo (1996)
(ii) without electroporation	Zebrafish	1 week	38	Khoo et al. (1992)
	Common carp	1-2 week		Muller et al. (1992)

	Rainbow trout	7 week		Chourrout and Perrot (1992)
	Indian catfish	Fry		Marian et al. (1997)
	Loach	Fry	*5	Zelenin et al. (1991)
4.Gene bombardment	Zebrafish	Fry	5	Zelenin et al. (1991)
	Rainbow trout	Fry		Zelenin et al. (1991)
	Medaka	Fry	13	Yamauchi et al. (2000)

*- Genomic integration reported #- Cumulative value of cytoplasmic persistence and genomic integration

common carp, zebrafish and channel catfish, in terms of gene transfer efficiency. Hence, electroporation is considered as an effective and versatile method for massive gene transfer (Powers et al., 1992; Chen et al., 1996), especially in fish, wherein microinjection is difficult due to invisibility of nucleus, tough chorion, etc. (Pandian and Marian, 1994). The eggs of temperate fish such as salmon and trout commence cell division several hours after fertilization, i.e. they have a long time window for microinjection (Chen et al., 1996). However, the eggs of tropical fish such as Indian carps commence cell division less than 20 to 30 minutes after fertilization and, thus, only have a shorter time window for microinjection. Hence, electroporation is the preferred method for gene transfer in tropical fish such as major Indian carps.

The technique of transferring the transgene into rohu eggs using electroporation encounters an unusual problem. Since the fertilized rohu eggs are large (1.8 ± 0.1 mm), the largest cuvette of the Gene pulser II (0.4 cm gap, catl no: 165-2088 of Biorad) can accommodate a maximum of 150 eggs and buffer. Moreover, the rohu eggs are fragile and swell by 2-3 times within 15-30 minutes following fertilization and attain a size of 2.5 to 5.0 mm in diameter. When ~150 eggs are accommodated in the cuvette for electroporation, ≥ 95% of the eggs suffer mortality (even at 150 Volt/cm) due to swelling (Venugopal et al., 1998b). Hence, only about 50 eggs can be electroporated at a given time. However, this problem has not been reported in the electroporation of the egg in the European common carp *Cyprinus carpio* (Powers et al., 1992), indicating the peculiar nature of the rohu eggs. Consequently, there was a need for a standardizing protocol for electroporetic transfer of transgene into the sperm for fertilizing normal eggs. This unusual situation led to the development of protocols for sperm-mediated gene transfer into the eggs of all the three major Indian carps (Venugopal et al., 1998b). Among all the studies, in which sperm electroporation was employed as gene transfer method, the current study is the first, wherein the transgene was traced for a period longer than 60 weeks (Table 1.5).

To evolve the 'best method' for gene transfer in fish, researchers (Inoue et al., 1992; Zhu et al., 1992; Powers et al., 1992; Kavumpurath et al., 1993; Marian, 1995) have made comparative studies on different methods of gene transfer in the species of their choice. Whereas the first three reviews do not proclaim any one of the gene transfer methods as 'best', Marian (1995) suggested that sperm electroporation is a better method, as it leads to higher percentage of integration, persistence and transmission—though extrachromosomally—in the Indian catfish. Powers et al., (1992) have reported that electroporation resulted in significantly greater numbers of transgenic fish than microinjection. The integration frequency was also much higher when electroporation was used for gene transfer in common carp and channel catfish, when compared to microinjection (Table 1.4).

Table 1.4 lists the use of different methods for gene transfer along with the transfer frequency. An analysis of the data indicates that microinjection and electroporation led to about 40% to 55% frequency of transgene, which persisted for longer than 1 year in rainbow trout, Atlantic salmon and nigrobuna, when transferred by microinjection, and for a period of 1 year in the Indian catfish (Sheela et al., 1999), when transferred by electroporation. Clearly, both microinjection and electroporation of the fertilized eggs appear to be equally efficient in ensuring persistence of the transgene in fairly high frequency and for-long periods. Hence, it appears that the chance factor plays a more crucial role in integration and persistence of the transgene, which will decrease considerably when sperm is used as a carrier with/without electroporation. However, there are a number of marine demersal fish, whose eggs are known to swell following fertilization (Fluchter and Pandian, 1968). Therefore, one has to use the sequences like 'eukaryotic origin of replication' to facilitate persistence and replication of the transgene, as has been done in the present study.

Table 1.5 : Transgenes used for gene transfer in fish by sperm electroporation. Age refers to the maximum period up to which the transgene was traced

Transgene	*Fish*	*Age*	Reference
pRSVCAT, pCMVNEO	African catfish, carp, tilapia	10 days	Muller et al. (1992)
pRSVCAT	rainbow trout	7 weeks	Chourrout and Perrot (1992)
pCMVlacz	salmon	Fry	Sin et al. (1993)
pCMVlacz, pRSVCAT	salmon	3 weeks	Walker et al. (1995)
pMTL	zebrafish	2 weeks	Patil and Khoo (1996)
pAFPsGH	loach	2 weeks	Tsai et al. (1995)
pRSVLTRrtGH	Indian major carps	2-3 weeks	Venugopal et al. (1998)
pCMV-rGH-IRES2-EGFP	rohu	60 weeks	Venugopal et al.(2004)

Table 1.5 shows that in the presence of the replication origin, the transgene persists for longer than 60 weeks, while it persists for just less than 3 weeks in the transgenic fish that contained the vector, perhaps without the replicon (Tsai et al., 1995; Venugopal et al., 1998b).

Eukaryotic replicon

The 'origin of replication' is essential for the perpetuation of any DNA in vivo (Darnell et al., 1986), and the plasmids that are lacking the origin of replication will be diluted out of their host cells (Old and Primrose, 1985) in due course of time. Unfortunately, several researchers have not realized the importance of this theory and so have not included eukaryotic replicons in their vectors (e.g. Dunham et al., 1987; however, see Alestrom et al., 2004). Others, who have made vectors exclusively for fish, have not emphasized its importance at all (e.g. Liu et al., 1990; Cavari et al., 1993; Takagi et al., 1994). Most of the investigators have not given any details about the replicon, although some investigators have used only the purified coding sequence along with the promoter, excluding the plasmid backbone (Brem et al., 1988; Nam et al., 2001).

Nam et al., (1999) have suggested that the extrachromosomally persisting transgene might be degraded, diluted or lost during the development of the fish. Conversely, Venugopal et al. (2004) have shown that the transgene persists for longer than a year. This finding supports the inclusion of an eukaryotic replicon to ensure the replication and maintenance of the transgene in the host, even when the transgene is not integrated. Experimental proofs are available to support the persistence of transgenic episome in germ cells (Hackett, 1994) and its transfer to F_1 progenies (Stuart et al. 1988). Obviously,

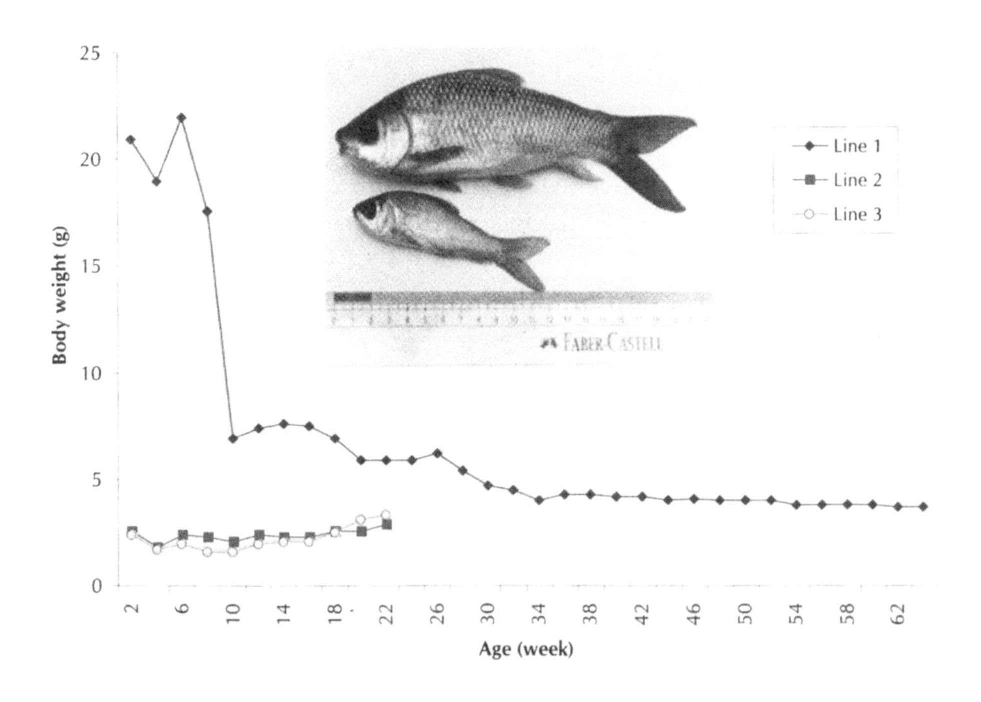

Fig. 1.1: Changes in growth acceleration as function of age in 3 different family lines of transgenic rohu. Window insert: the largest transgenic is shown along with control.

extrachromosomal persistence may not be a limiting factor for the perpetuation of a transgene and its transmission to the transgenic lines, provided the vector is endowed with a facility for self replication. However, the mechanism that results in persistence of the gene as extrachromosomal entity is not clear. Rokkones et al., (1989) have suggested the formation of micronucleus that are protected from nucleases and replicated by the host cell polymerases. Formation of 'nucleoid-like' structures as a result of chromatin and nuclear assembly of protein has been reported in *Xenopus* (Forbes et al., 1983; Newport, 1987; Lenow and Lasky, 1991) and the formation of such structures in fish remains a possibility that has not yet been explored (Iyengar et al. 1996).

GROWTH ACCELERATION IN AUTOTRANSGENIC *LABEO ROHITA*

Three lines of fast-growing autotransgenic *Labeo rohita* were generated with homologous growth hormone cDNA containing construct (pCMV-rGH-IRES2-EGFP) by electroporated gene mediated transfer. Both control and experiment rohu suffered 99% mortality within the first 10 weeks of hatching, but none suffered any deformity. Putative transgenic belonging to the family lines 1 and 2 consistently grew 2-3 times faster than their respective siblings, but all of them succumbed within 24 weeks. Transgenics belonging to the family line 1 displayed growth acceleration more than 20 times initially, but the level of acceleration progressively decreased to 3.7 times at the age of 64 weeks (Fig. 1.1). Clearly, there were remarkable differences in the expression of growth acceleration among the putative transgenics belonging to different family lines. The sharp decline in mean growth acceleration at the tenth week was also synchronized with mortality of more than 2,400 fingerlings, most of which had displayed higher growth acceleration. Fast-growing transgenics appeared to be more vulnerable than the others. However, slot blot analysis showed that about 25% of the total fry belonging to all the 3 lines were fin-positive at the age of 3 weeks. Slot and Southern analyses confirmed the consistent presence of the transgene at the same frequency at the same individual until the age of 60 weeks. The analyses also conformed the fact that the introduced transgene remained absolutely intact, but only extrachromosomally. Southern analysis of the fin-positive transgenics indicated that selected organs arising from the germinal layers were also positive. RT-PCR analysis of RNA extracted from blood of the fin-positive transgenics showed that 5 out of 6 were positive (see Pandian, 2002).

Figure 1.2 displays the levels and trends obtained for growth acceleration in selected species of transgenic fish. Many authors have provided only one value (e.g. loach: Tsai et al., 1995; tilapia: Rahman et al., 1998); others have given more values and extended to F_1 (e.g. zebra fish: Pandian et al.,1991; catfish: Sheela et al., 1999), F_2 and F_3 progenies (mudloach: Nam et al., 2001). These reported values for growth acceleration of the generated transgenic fish suggest 3 different trends: (i) stable growth acceleration at levels $\approx$ 1.2 times (zebrafish: Pandian et al., 1991), $\approx$ 3 times (catfish: Sheela et al., 1999; rohu family line 2 and 3), and ≈ 19 times (mudloach, Nam et al., 2001); (ii) an increasing trend from 1.8- 2.4 times (loach: Tsai 1995) and from 2.8 to 5 times (Atlantic salmon: Du et al., 1992) and (iii) a steep fall from about 20 times and subsequent stabilization at about 5 times (line 1 rohu; this study). Remarkably, all these transgenic fish have displayed definite and measurable growth acceleration. However, it is difficult to understand why the levels and trends of growth hormone transgenics differ so widely.

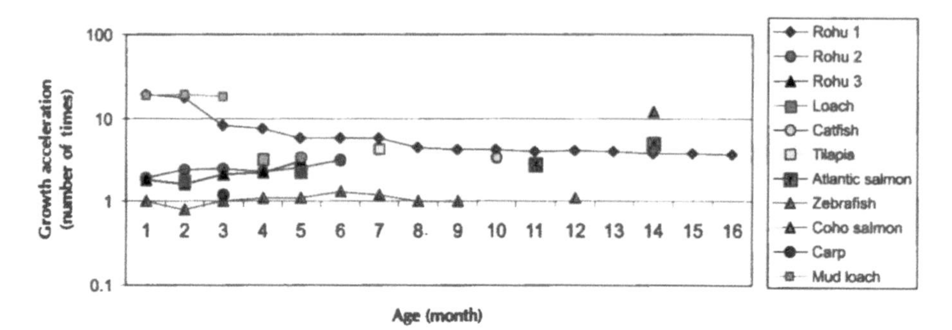

Fig. 1.2: Trends obtained for growth acceleration in selected transgenic fish

ACKNOWLEDGEMENTS

Financial assistance provided by the Council of Scientific and Industrial Research, Indian Council of Agriculture Research and Department of Biotechnology, New Delhi is gratefully acknowledged.

REFERENCES

Alam, M.S., Popplewell, A. and Maclean, N. 1995. Germ line transmission and expression of a lacZ containing transgene in tilapia (*Oreochromis mossambicus*). Transgenic Res., 4: 51-56.

Alestrom, P., Klungland. H., Kisen. G. and Anderson, O. 1991. Fish GnRH gene and molecular approaches for control of sexual maturation. Abstracts S75. p. 65. In: Programs and Abstracts, IMBC 13-16 October, Baltimore, USA.

Anathy, V., Venugopal, T., Koteeswaran, R., Pandian, T.J., and Mathavan, S. 2001. Cloning, sequencing and expression of cDNA-encoding growth hormone from Indian catfish (*Heteropneustes fossilis*). J. Biosci., 26: 315-324.

Bernardi, G., D'onotrio, G., Caccio, S. and Bernardi, G. 1993. Molecular phylogeny of bony fishes based on their amino acid sequences of the growth hormone. J. Mol. Evol., 16, 644-649.

Biswas, M., Kanapin, A. and Apweiler, R. 2001. Application of Interpro for the functional classification of the proteins of fish origin in Swiss-Prot and TrEMBL. J. Biosci., 26: 277-284.

Brem, G., Brenig, B., Horstgen-Schwark, G. and Winnacker, E.L. 1988. Gene transfer in tilapia (*Oreochromis niloticus*). Aquaculture, 68: 209-219.

Brenner, S., Elgar, G., Sandfors, R., Macrae, A., Venkatesh B. and Aparicio, S. 1993. Characterization of pufferfish (fugu) genome as a compact model vertebrate genome. Nature, 366: 265-268.

Buno, R.J. and Linser, P.J. 1992. Transient expression of pRSVCAT in transgenic zebrafish made by elctroporation. Mol. Mar. Biol. Biotechnol., 1: 271-275.

Cavari, B., Funkenstein, B. and Chen, T.T. 1993. Effect of growth hormone on the growth rate of the gilthead seabream (*Sparus aurata*) and the use of different constructs for production of transgenic fish. Aquaculture, 111: 189-197.

Chan, W.K. and Devlin, R.H. 1993. Polymerase chain reaction amplification and functional characterization of sockeye salmon histone H3, metallothionein-b, and protamine promoters. Mol. Mar. Biol. Biotechnol., 2: 308-318.

Chang, Y.S., Liu, C.S., Huang, F.L. and Lo, T.B. 1992. The primary structures of growth hormones of three cyprinid species: Bighead carp, Silver carp and Grass carp; Gen. Comp. Endocrinol., 87: 385-393.

Chen, T.T., Knight, K. and Lin, C.M. 1993. Expression and inheritance of RSV-LTR-rtGH1 cDNA in the transgenic common carp, *Cyprinus carpio*. Mol. Mar. Biol. Biotechnol., 2: 88-95.

Chen, T.T., Vrolijik, N.H., Lu, J.K., Lin, C.M., Reimschuessel, R. and Dunham, R.A. 1996. Transgenic fish and its application in basic and applied research. Ann. Rev. Biotechnol., 1: 205-237.

Chiou, C.S., Chen, H.T. and Chang, W.C. 1990. The complete nucleotide sequence of the growth hormone gene from common carp (*C. carpio*). Biochem. Biophys. Acta, 1087: 91-94.

Chong, S.S.C. and Vielkind, J.R 1989. Single step and fate of CAT reporter gene-microinjected into fertilized-medaka (*Oryzias latipes*) eggs in the form of plasmid. Theor. Appl. Genet., 78:369–380.

Chourrout, D.R., Guyomard, R. and Houdebine, L.M. 1986. High efficiency gene transfer in rainbow trout (*Salmo gairdneri Rich.*) by microinjection into egg cytoplasm. Aquaculture, 51: 143-150.

Chourrout, D. and Perrot, E. 1992. No transgenic rainbow trout produced with sperm incubated with linear DNA. Mol. Biol. Biotechnol., 1: 257-265.

Church, R.B. 1987. Embryo manipulation and gene transfer in domestic animals. Trends Biotechnol., 5: 13-19.

Clark, A.J., Simons, P., Wilmut, I. and Lathe, R. 1987. Pharmaceuticals from transgeneic livestock. Trends Biotechnol., 5: 20-24.

Collas, P. and Alestrom, P. 1998. Nuclear localization signals enhance germline transmission of a transgene in zebrafish. Transgenic Res., 7: 303-309.

Darnell, J., Lodish, H. and Baltimore, D. 1986. Molecular Cell Biology, Scientific American Books Inc., New York pp 433- 438.

Depoint- Zilli, L., Seiler-Tuyns, A. and Paterson, B.M. 1988. A 40 base pair sequence in the 3' end of the β-actin mRNA transcription during myogenesis. Proc. Natl. Acad. Sci., USA, 85: 1389-1393.

Devlin, R.H., Yesaki, T.Y., Blagl, C.A. and Donaldson, E.M. 1994. Extraordinary salmon growth. Nature, 371: 209-210.

Du, S.J., Gong, Z., Fletcher, G.L., Shears, M.A., King, M.J., Idler, D.R. and Hew, C.L. 1992. Growth enhancement in transgenic Atlantic salmon by the use of an 'all-fish' chimeric growth hormone gene construct. Biotechnol., 10: 176-181.

Dunham, R.A., Eash, J., Askins, J. and Towers, T.M. 1987. Transfer of metallothionein human growth hormone fusion gene into channel catfish. Trans. Amer. Fish. Soci., 116: 87-91.

Elgar, G., Sandford, R., Aparicio, S., Macrae, A., Venkatesh, B. and Brenner, S., 1996. Small is beautiful: Comparative genomics with the pufferfish (*Fugu rubripes*). Trends. Genet., 12:145-149.

Erdelyi, F. Papp. T., Muller, F., Adam, A., Egedi, S., Stetak, A. and Orbanm, L. 1994. Production of transgenic catfish and carp by co-injecting a reporter gene with growth hormone gene. Proc. 3rd Internatl. Mar. Biotechnol. Conf. (IMBC, 1994), p. 71.

Etkin, L.D., Pearman, B., Robeils, M. and Bektesh, S.L. 1984. Replication, integration and expression of exogenous DNA injected into fertilized eggs of *Xenopus laevis*. Differentiation, 2: 194-202.

Evans, H.M. and Long, J.H. 1921. The effect of anterior lobe administered intraperitoneally upon growth, maturity and olfactory cycles of rat. Anat. Rec., 21: 62-63.
Fahrenkrug, S.C., Clark, K.J., Dahlquist, M.O and Hackett, P.B. Jr. 1999. Bicistronic gene expression in developing zebrafish. Mar. Biotechnol., 1: 552–561
Fletcher, G.L. Shears, M.A., King, M.J., Davies, P.L and Hew, C.L. 1998. Evidence for antifreeze protein gene transfer in Atlantic salmon (*Salmo salar*). Can. J. Fish. Aquat. Sci., 45: 352-357.
Fluchter, J. and Pandian, T.J. 1968. Rate and efficiency of yolk utilization of eggs of the sole *Solea solea*. Helgolander wiss. Meeresunters., 18: 53–60.
Forbes, D.J., Kirschner, M.N. and Newport, J.N. 1983. Spontaneous formation of nucleus-like structures around bacteriophage lambda DNA. Cell Biology, 34: 13-23.
Friedenreich, H. and Schartl, M. 1990. Transient expression directed by homologous and heterologous promoter enhancer sequences in fish cells. Nucleic Acids Res., 18: 3299-3305.
Funkenstein, B., Cavari, B., Moav, B., Harari, O. and Chen, T.T. 1991. Gene transfer of growth hormone in the Gilthead sea bream (*Sparus aurata*) and characterization of its pregrowth hormone cDNA. 2nd Internatl. Mar. Biotechnol. Conf. (IMCB 1991). p.63.
Gong, Z., Du, S.J., Fletcher, G.L., Shears, M.A., Davies, P.L., Devlin, B. and Hew, C.L 1994. Biotechnology for aquaculture: Transgenic salmon with enhanced growth and freeze-resistance. Proc. Internatl. Symp. Biotechnol. Appl. Aquacult. p. 83.
Gross. M.L., Schneider, J.F., Moav, N., Alvarex, C., Myster, S., Liu, Z., Hew, C.L., Hallerman, E.M., Hackett, P.B., Guise, K.S., Faras, A.J. and Kapuscinski, A.R. 1991. Molecular analysis and growth evaluation of northern pike (*Esox lucius*) microinjected with growth hormone genes. Aquaculture, 85: 115-128
Hackett, P.B., Caldovic, L., Izsvak, Z., Ivics, Z., Fahrenkrug, S., Kaufman, C.M., Martinez, G. and Essner, J.E. 1994. Production of transgenic fish. Proc. 3rd Internatl. Mar. Biotechnol. Conf. (IMBC, 1994), p 73.
Hammer, R.E., Pursel, V.G., Rexroad, C.E., Wall, R.J., Ebert, K.M., Palmiter, R.D. and Brinster, R.L. 1985. Production of transgenic rabbits, sheep and pigs by microinjection. Nature, 315: 680-683.
Hayat, M., Joyce, C.P., Townes, T.M., Chen, T.T., Powers, D.A. and Dunham, R.A. 1991. Survival and integration rate of channel catfish and common carp embryos microinjected with DNA at various developmental stages. Aquaculture, 99: 249-256.
Hellen, C.U.T. and Sarnow, P. 2001. Internal ribosome entry site in eukaryotic mRNA molecules. Genes Dev., 15: 1593-1612.
Hew, C.L. 1989. Transgenic fish: Present status and future directions. Fish. Physiol. Biochem., 7: 409-413.
Hong, Y., Winkler, C., Brem, G. and Schartl, M. 1993. Development of a heavy metal-inducible fish-specific expression vector for gene transgenic in vitro and in vivo. Aquaculture, 111: 215-226.
Inoue, K., Yamashita, S., Hata, J., Kabeno, S., Asada, S., Nagashisa, E. and Fujita, T. 1990. Electroporation as a new technique for producing transgenic fish. Cell Differ. Dev., 29: 123-128.
Ivics, Z., Izavak, Z. and Hackett, P.B. 1993. Enhanced incorporation of transgenic DNA into zebrafish chromosome by a retrovirus integration protein. Mol. Mar. Biol. Biotechnol., 2: 162-173.
Iyengar, A., Muller, F. and MaClean, N. 1996. Regulation and expression of transgenic fish—A review. Transgenic Res., 5: 147-166.
Kavumpurath, S., Anderson, O., Kisen, G. and Alestrom, P. 1993. Gene transfer methods and luciferase gene expression in zebrafish, *Brachydanio rerio* (Hamilton). Israeli J. Aquat., 45: 154-163.
Khoo, H.W., Ang, L.H., Lim, H.B. and Wong, K.Y. 1992. Sperm cells as vectors for introducing foreign DNA into zebrafish. Aquaculture, 107: 1-19.
Le Gac, F., Blaise, O., Fostier, A., Le Bail, P.Y., Loir, M., Mourot, B. and Weil, C. 1993. Growth hormone (GH) and reproduction: A review. Fish Physiol. Biochem., 11: 219-232.

Lenow, G.H. and Laskey, R.A. 1991. The nuclear membrane determines the timing of DNA replication in *Xenopus* egg extracts. J. Cell Biol., 112: 557-566.

Li, C.H. and Evans, H.M. 1944. The isolation of pituitary growth hormone. Science, 99: 183-84.

Liang, M.R., Alestrom, P. and Collas, P. 2000. Glowing zebrafish: Integration, transmission and expression of a single luciferase transgene promoted by non-covalent DNA nuclear transport peptide complex. Transgenic Res., 55: 8-13.

Liu, Z., Moav, B., Faraz, A.J., Guise, K.S., Kapuscienski, A.R. and Hackett, P.B. 1990. Development of expression vectors for transgenic fish. Biotechnology, 18: 1268-1272.

MaClean, N., Iyengar, A., Rahman, A., Sulaiman, Z. and Penman, D. 1992. Transgene transmission and expression in rainbow trout and tilapia. Mol. Mar. Biol. Biotechnol., 1: 355-365

MaMahan, A.P., Novak, J.J., Britten, R.J. and Davidson, E.H. 1984. Inducible expression of a cloned heat shock fusion gene in sea urchin embryos. Proc. Natl. Acad. Sci. USA, 81: 7490–7494.

Marian, L.A. 1995. Ploidy induction and genetic studies in ornamental and food fish. Doctoral dissertation. Madurai Kamaraj University, Madurai, India.

Marian, L.A., Pandian, T.J., and Mathavan, S. 1997. Gene transfer, integration, expression and transmission of rainbow trout growth hormone gene in the Indian catfish: A comparative study. Proc. International Symposium on Marine Biotechnology. Phuket, Thailand.

Martiniez-Salas. H.M., 1999. Internal ribosome entry site biology and its use in expression vectors. Curr. Opin. Biotechnol., 10: 458-464.

Mathews, L., Norstedt, G. and Palmiter., R.D. 1986. Regulation of insulin like growth factor I gene expression by growth hormone. Proc. Natl. Acad. Sci. USA, 83: 9343-9347.

Miller, W.L. and Eberhardt, N.L. 1983. Structure and evolution of the growth hormone gene family. Endrocr. Rev., 4: 97-130.

Moav, B., Hinitis, Y., Gnoll, Y. and Rothbard, S. 1995. Inheritance of recombinant carp beta-actin GH cDNA in transgenic carp. Aquaculture, 137: 179-185

Mountford, P.S., and Smith, A.G. 1995. Internal ribosome entry sites and dicistronic RNAs in mammalian transgenesis. Trends. Genet., 4: 179-84.

Muller, F., Ivics, Z., Eredelyi, F., Vardi, L., Horvath, L., Maclean, N. and Orban, L. 1992. Introducing foreign genes into fish eggs by electroporated sperm as a carrier. Mol. Mar. Biol. Biotechnol., 1: 276-281.

Nam, Y.K., Noh, C.H. and Kim, D.S. 1999. Transmission and expression of an integrated reporter construct in three generations of transgenic mudloach (*Misgurnus mizolepis*). Aquaculture, 172: 229-245.

Nam, Y.K., Cho, Y.S., Chang, Y.J., Jo, J.Y. and Kim, D.S. 2000. Generation of transgenic homogenous lines carrying CAT gene in mudloach. Fish. Sci., 66: 58-82.

Nam, Y.K., Noh, J.K., Cho, Y.S., Cho, H.J., Cho, K.N., Kim, C.G. and Kim, D.S. 2001. Dramatically accelerated growth and extraordinary gigantism of transgenic mudloach (*Misgurnus mizolepis*). Transgenic Res., 10: 353-362.

Newport, J. 1987. Nuclear reconstitution in vitro: Stages of assembly around protein-free DNA. Cell, 48: 205-217.

Nial, H.D., Hogan, M.L. and Saver, R. 1971. Sequence of pituitary and placental lactogenic and growth hormones: Evaluation from a primordial peptide by gene reduplication. Proc. Natl. Acad. Sci. USA, 66: 866-870.

Old, R.W. and Primrose, S.B., 1985. Principles of Gene Manipulation, Blackwell Scientific Publishers, Oxford pp 3-5.

Ozato, K., Kondoh, H., Inohara, H., Iwamatsu, T., Wakamatsu, Y. and Okada, T.S. 1986. Production of transgenic fish: Introduction and expression of chicken δ-crystalline gene in medaka embryos. Cell Differ. Dev., 19: 234-237.

Palmiter, R.D., Brinster, R.L., Hammer, R.E., Trumbauer, M.E., Rosenfeld, M.G., Birnberg, N.C. and Evans, R.M. 1982. Dramatic growth of mice that develop from eggs microinjected with metallothionein-growth hormone fusion genes. Science, 300: 611-615.

Palmiter, R.D. and Brinster, R.L. 1986. Germline transformation of mice. Ann. Rev. Genet., 20: 465-49.

Pandian, T.J. 1995. Transgenic animals. Monograph. Pondichery University p. 1-12.

Pandian, T.J. 2002 Transgenesis in fish: Indian endeavour and acheivement J. Aquacult., 6: 51-58

Pandian, T.J. and Marian, L.A. 1994. Problems and prospects of transgenic fish production. Curr. Sci., 66: 635-649.

Pandian, T.J. and Sheela, S.G. 1995. Hormonal induction of sex reversal. Aquaculture, 130: 1-22.

Pandian, T.J. and Koteeswaran, R. 1998. Ploidy induction and sex control in fish. Hydrobiologia, 384: 167-243.

Pandian, T.J., Mathavan, S. and Marian, L.A. 1994. Need for genetic and molecular biological research in Indian fish. Curr. Sci., 66: 633-634.

Pandian, T.J., Venugopal, T. and Koteeswaran, R. 1999. Problems and prospects of hormone, chromosome and gene manipulations in fish. Curr. Sci., 76: 369-386.

Pandian, T.J., Kavumpurath, S., Mathavan, S. and Dharmalingam, K. 1991. Microinjection of rat growth hormone gene into zebrafish eggs and production of transgenic zebrafish. Curr. Sci., 60: 596-600.

Patil, G. and Khoo, H.W. 1996. Nuclear internalization of foreign DNA by zebrafish spermatozoa and its enhancement by electroporation. J. Exp. Zool., 274: 121-129.

Pitkanen, T.I., Krasanov, A., Tteerijioki, H. and Molsa, M. 1999. Transfer of growth hormone (GH) transgenes into Arctic charr (*Salvelinus alpinus*): Growth response to various GH constructs. Genet. Analy. Biomol. Eng., 15: 91-98.

Postlewait, J.H. and Talfot, W.S. 1997. Zebrafish genomics: From mutants to genes. Trends. Genet, 13: 183-190.

Powers, D.A., Hereford, L., Cole, T., Chen, T.T., Lin, C.M., Knight, K., Creech, K. and Dunham, R. 1992. Electroporation: A method for transferring genes into the gametes of zebrafish (*Brachydanio rerio*), channel catfish (*Ictalurus punctatus*) and common carp (*Cyprinus carpio*). Mol. Mar. Biol. Biotechnol., 1: 301-308.

Rahman, M.A., Hwang, G.L., Razaka, S.A., Sohm, F. and Maclean, N. 2000. Copy number related transgene expression and mosaic somatic expression in hemizygous and homozygous transgenic tilapia (*Oreochromis niloticus*). Transgenic Res., 9: 417-427.

Rahman, M.A., Mak, R., Ayad, H., Smith, A. and Maclean, N. 1998. Expression of a novel piscine growth hormone gene results in growth enhancement in transgenic tilapia (*Oreochromis niloticus*). Transgenic Res., 7: 357-369.

Ralph, R.K., Darkin, O., Rattray, S. and Schofield. P., 1990. Growth-related protein kinases. Bioessays, 12: 121–124.

Reddy, P.V.G.K., Tripathi, S.D., Jana, R.K., Mahapatra, K.D., Saha, J.N., Meher, P.K. Ayyappan, S., Sahoo, M., Lenka, S., Govindasamy, P., Gjerde, B. and Rye, M. 2001. Project report on selective breeding of rohu for sustainable aquaculture. Central institute of Freshwater Aquaculture (CIFA) Kausalyaganga, Orissa, India.

Rokkones, E.P., Alestrom, P., Skjervold, M., Gautvik, K.M. 1989. Microinjection and expression of a mouse metallothionein human growth hormone fusion gene in fertilized salmonid eggs. J. Comp. Physiol., 158: 751-758.

Rublin, G.M., Spradling, A.C. 1982. Genetic transformation of *Drosophila* with transposable element vectors. Science, 1218: 348 -353.

Sakamoto, T., Mc Cormick, S.D. and Hirano, T. 1993. Osmoregulatory actions of growth hormone and its mode of action in salmonids: A review. Fish. Physiol. Biochem., 11: 155-164.

Sara, V.R. 1991. Insulin like growth factor and fetal growth. In: Encyclopedia of Human Biology, Academic Press, New York. Vol.4: pp. 503-511

Sheela, S.G., J.D., Chen, Mathavan, S. and T.J. Pandian, 1998. Construction, electroporetic transfer and expression of ZpbypGH ZpbrtGH in zebrafish. J. Biosci., 23: 565-576.

Sheela, S.G., Chen, J.D., Mathavan, S. and Pandian, T.J. 1999. Electroporetic transfer stable integration, expression of ZpbypGH ZpbrtGH in Indian catfish *Heteropneustes fossilis*. Aquacult. Res., 30: 233-248.

Sheela, S., 2003. Gene transfer studies in fishes. Ph.D thesis, Kerala University, Thiruvananthapuram.

Sin, F.Y.T., Bartely, A.L., Walker, S.P., Sin, I.L., Symonds, J.E. Howke, L. and Hopkino, C.C. 1993. Gene transfer in chinook salmon (*Oncorhynchus tshawytscha*) by electroporesis sperm in the presence of pRSV-lacz DNA. Aquaculture, 117: 57-69.

Stinchcomb, D.T., Shaw, J.E., Carr, S.H., and Hirsh, D. 1985. Extrachromosmoal DNA transformation of *Caenorhabditis elegans*. Mol. Cell Biol., 5: 3484-3496.

Stuart, G.W., McMurray, J.V. and Westerfield, M. 1988. Replication, integration and stable germ-line transmission of foreign sequences injected into early zebrafish embryos. Development, 103: 403-412.

Takagi, S., Sasado, T., Tamiya, G., Ozato, K., Wakamatsu, Y. Takeshita, A. and Kimura, M. 1994. An efficient expression for transgenic medaka construction. Mol. Mar. Biol. Biotechnol., 3: 192-199.

Tamiya, E., Sugiyama, T., Masaki, K., Hirose, A., Okoshi, T. and Karube, I. 1990. Spatial imaging of luciferase gene expression in transgenic fish. Nucleic Acids Res., 18: 1072.

Tsai, H.J., Tseng, F.S. and Liao, I. 1995. Electroporation of sperm to introduce foreign DNA into the genome of loach (*Misgurnus anguillicaudatus*). Can. J. Fish Aquat. Sci., 52: 776-778.

Tseng, F.S., Liao, I.C. and Tsai, H.J. 1994. Introducing the exogenous growth hormone cDNA into loach (*Misgurnus anguillicaudatus*) egg via the electroporated sperms as carrier. Proc. 3rd Internatl. Mar. Biotechnol. Conf. (IMBC 1994), p. 71.

Ueno, K., Hamaguichi, S., Ozato, K., Kang, J.H. and Inoue, K. 1994. Foreign gene transfer into nigrobuna (*Carassius auratus grandoculis*). Mol. Mar. Biol. Biotechnol., 3: 235-342.

Van der Kraak, G., Rosenblum, P.M. and Peter, R.E. 1990. Growth hormone dependent potentiation of gonodotropin stimulated steroid production by ovarian follicles of the gold fish. Gen. Comp. Endocrinol., 71: 233-239

Venugopal, T., Mathavan, S. and Pandian, T.J. 1998a. Cloning, sequencing and comparison of growth hormone cDNA of Indian major carps. Proc. Fifth Asian Fisheries Forum, Thailand, p. 136.

Venugopal, T., Pandian, T.J, Mathavan, S. and Sarangi, N. 1998b. Gene transfer in Indian major carps by electroporation. Cur. Sci., 74: 636-640.

Venugopal, T., Anathy, V., Pandian, T.J. and Mathavan, S. 2002. Molecular cloning of growth hormone encoding cDNA of an Indian major carp, *Labeo rohita* and its expression in *E. coli* and zebrafish; Gen. Comp. Endocrinol., 125: 236-247.

Venugopal, T., Anathy, V. and Pandian, T.J., 2004. Growth enhancement and food conversion efficiency of transgenic fish *Labeo rohita*. J. Exp. Zool., 301A: 477-490.

Walker, S.T., Symonds, J.E., Sin, I.L., and Sin, F.Y.T. 1995. Gene transfer by elctroporated chinook salmon sperm. J. Mar. Biotechnol., 3: 232-234.

Wang, X., Korzh, V. and Gong, Z. 2000. Use of IRES bi-cistronic construct to trace expression of exogenously injected mRNA in zebrafish embryos. Biotechniques, 29: 814-819.

Winkler, C., Hong, Y., Wittbrdt, J. and Schartl, H.M. 1992. Analysis of heterologous and homologous enhancers in vivo and in vitro by gene transfer into Japanese medaka (*Oryzias latipes*) and *Xiphophorous*. Mol. Mar. Biotechnol., 1: 326-337.

Xie, Y., Liu, D., Zou, J., Li, G. and Zhu, Z. 1993a. Gene transfer via electroporation in fish. Aquaculture, 111: 207-213.

Xie, Y., Liu, D., Zou, J., Li, G. and Zhu, Z. 1993b. Novel gene transfer in the fertilized eggs of loach via electroporation. Acta Hydrobiol. Sinica., 13: 387-389.

Yamauchi, M., Kinoshita, M., Sosanuma, M.T., Suji, S., Terada, M., Morimyo, M. and Ishikawa, Y. 2000. Introduction of a foreign gene into medaka fish using the particle gun method. J. Exp. Zool., 287: 285-293.

Yoshizaki, G., Oshiro, T. and Takashima, F. 1991. Introduction of carp globin into rainbow trout. Nippon Suisan Gakkaishi, 57: 819-824.

Zelenin, A.V., Alimov, A.A., Barmintzov, A.O., Zeleina, I.A., Krasnov, A.M. and Kolesnikov, V.A. 1991. The delivery of foreign genes into fertilized fish eggs using high velocity microprojectiles. FEBS Lett., 287: 118-120.

Zhang, P., Hayat, M., Joyce, C., Gonzalez-Villasenor, L.I., Lin, C.M., Dunham, R.A., Chen,

T.T. and Powers, D.A. 1990. Gene transfer, expression and inheritance of pRSV-rainbow trout GHcDNA in the common carp, *Cyprinus carpio* (Linnaeus). Mol. Reprod. Dev., 25: 3-13.

Zhao, X., Zhang, P.T., and Wong, R. 1993. Gene transfer in goldfish, *Carassius auratus* by oocyte microinjection. Aquaculture, 111: 311.

Zhu, Z., Li, G., He, L. and Chen, S. 1985. Novel gene transfer into the fertilized eggs of goldfish (*Carassius auratus* L. 1758). J. App. Ichthyol., 1: 31-34.

Zhu, Z. 1992. Generation of fast growing transgenic fish; methods and mechanisms. In: Transgenic Fish, (eds)., Hew, C.L. and Fletcher, G.L., World Scientific Publishing, Singapore, p. 92–119.

Zhu, Z, and Sun, Y.H. 2000. Embryonic and genetic manipulation in fish. Cell Res., 10 : 17-27.

Chapter 2

Gene Transfer to Germline and Somatic Tissues of Zebrafish

P. Aleström, P. Collas,* J. Torgersen, M-r. Liang,.**
A. Arenal,*** and **R. Nourizadeh-Lillabadi**
Department of Basic Sciences and Aquatic Medicine
Norwegian School of Veterinary Science,
PO Box 8146 Dep., N-0033 Oslo, Norway
Email: peter.alestrom@veths.no
*Institute of Medical Biochemistry, University of Oslo,
PO Box 1112 Blindern, 0317 Oslo, Norway
**Loeb Health Research Institute, Ottawa, Canada.
*** Center for Genetic Engineering and Biotechnology,
PO Box 387, 70100 Camaguey 1, Cuba

ABSTRACT

The development of vehicles driving foreign DNA into the cell nucleus is essential for effective gene transfer applications. We have previously reported that the use of Nuclear Localization Signal (NLS) peptides non-covalently complexed to plasmid DNA promotes the nuclear uptake of DNA and results in increased integration and germline transmission frequencies of the transgene. To achieve germline integration, inheritance and expression of single copy luciferase reporter gene in zebrafish, 10, 100 or 1000 copies of plasmid DNA complexed with NLS were used. As little as 10 DNA-NLS complexes (0.06 fg plasmid DNA) are adequate to produce germline-transgenic zebrafish bearing a single copy of the transgene. DNA transferred to zebrafish (*Danio rerio*) skin by gene gun revealed that already one hour after biolistic transfer of 1 µg reporter gene construct to zebrafish skin, 100% of the cells in the target area exhibited GFP gene activity. The risk for accidental germline uptake of DNA molecules delivered to dermis by a gene gun was shown to be below the detection level in a screen of 250,000 zebrafish sperm cells investigated by fluorescence in situ hybridization (FISH) at 15 or 30 days post delivery of 1 µg plasmid vector.

Keywords: Nuclear localization, Signal Peptides, gene-gun, accidental germline uptake of DNA

INTRODUCTION

In future food production, an increasing amount will be supplied by aquaculture. To manage this expected development, application of molecular biology tools will be instrumental in fighting the disease, production and environmental problems of intensive animal culture (Aleström, 1996). A key to strategies for efficient DNA immunization, gene therapy and most transgenesis applications lies in efficient uptake of foreign genes by the cell nucleus, where transient expression and/or genome integration occurs.

To date, the method of choice to generate transgenic fish consists in "forcing" a DNA construct into the nucleus of a fertilized egg by microinjecting large numbers of DNA (10^5-10^7 copies of plasmid DNA (see Pandian and Marian, 1994), which is equivalent to the amount of DNA in the fish genome into the cytoplasm (Iyengar et al., 1996). Under these conditions, the frequency of genomic integration and germline transmission of transgenes is usually in the order of a few percent (Stuart et al., 1988, 1990; Culp et al., 1991; Lin et al., 1994; Collas and Aleström, 1998). Recent studies have shown that a limiting step in fish transgenesis resides in slow nuclear import of DNA (Collas and Aleström, 1997, 1998, see also Pandian and Venugopal, 2004), a situation likely to favor late and hence inefficient and mosaic transgene integration in the germline of founder individuals (Stuart et al., 1988, 1990; Collas and Aleström, 1998; Culp et al., 1991).

Improvements in nuclear uptake of DNA have resulted from the use of protein-DNA complexes. Non-covalent attachment of DNA to karyophilic proteins has been shown to enhance nuclear import and expression of DNA in cultured mammalian cells (Kaneda et al., 1989; Fritz et al., 1996). Non-covalent (ionic) binding of plasmid DNA to synthetic nuclear localization signal (NLS) peptides analogous to NLS sequence of SV40 large T antigen also promotes nuclear import of DNA in vitro (Collas and Aleström, 1996) and in zebrafish embryos in vivo (Collas and Aleström, 1997). NLS-mediated nuclear import of DNA occurs through the nuclear pore system and exhibits biochemical requirements similar to nuclear protein import (Collas and Aleström, 1996). As a result of improved nuclear uptake of DNA-NLS complexes, enhanced expression and germline integration of a reporter transgene have been reported in zebrafish (Collas and Aleström, 1996, 1998). Covalent cross-linking of NLS peptides to DNA, unless terminal, will block transcription (Sebestyen et al., 1998; Zanta et al.,1999). As alternative to NLS peptides, a 372 bp SV40 NLS acting cis-element which contains the SV40 ORI and enhancer region, is shown to facilitate nuclear import and preferentially localize the plasmids to actively transcribed regions within the transfected nuclei of cells in culture (Dean, 1997). This is an additional important feature with respect to position effects.

Microinjection of high concentrations of DNA or DNA-NLS complexes into fish eggs almost inevitably results in the integration and inheritance of multiple copies of the transgene (Stuart et al., 1990; Collas and Aleström, 1998). Insertion of multiple transgene copies leads to increased risks of lethal integrations, and complex transgene expression and inheritance patterns.

The ability of naked DNA administrated through intra-muscular (i.m.) injection in order to cause a transient transformation of somatic cells (Wolff et al., 1990; see also Rolland, 1997) has opened the door for novel strategies for vaccination and gene therapy. The amount of DNA used for vaccination usually ranges from 1 to 1000 μg. The saline-based solute of DNA vaccines has no reported side effects in fish (Heppell et al., 1998). However, the efficiency of i.m. injections to fish may vary because of solute leakage from the site of injection (Heppell et al., 1998). Consequently, alternative DNA delivery systems have been investigated for their DNA transfer capability. A biolistic DNA transfer strategy has emerged with the hand-held Helios Gene Gun™ (Bio-Rad, Hercules, USA), which delivers microcarriers of gold coated with plasmid DNA through cell and nuclear membranes (Fynan et al., 1993; Gómez-Chiarri et al., 1996). In fish, the primary target tissues for biolistic mediated DNA vaccine delivery are epidermis, dermis and muscles.

Using luciferase and GFP reporter gene expression in zebrafish skin, we have studied the parameters influencing the efficiency of Gene gun-mediated DNA transfer to zebrafish epidermis and dermis. The results indicate that DNA transferred to zebrafish skin by Gene-gun is efficiently expressed, and that zebrafish may be used as a model organism for transient expression and DNA vaccines studies.

METHODS

Zebrafish and microinjection into zebrafish eggs

Methods for zebrafish care follow the protocols of the Zebrafish manual (Westerfield, 1993). Microinjection of plasmid DNA into fertilized zebrafish eggs is described in detail elsewhere (Collas et al. 1997).

Gene gun delivery of gold particles coated with plasmid DNA

The biolistic approach using gas pressure for accelerating DNA coated microcarriers (gold particles) was established for DNA transfer to skin (epidermis and dermis) of zebrafish using an Helios Gene Gun System (BIO-RAD) according to the protocol of the manufacturer (Torgersen et al. 2000). Optimal GFP reporter gene uptake and expression of both epidermis and dermis was achieved with 1 μm diameter microcarriers using a gas pressure of 200 psi.

Plasmid DNA–NLS peptide complexes

Plasmids used were pCMVL (luciferase reporter gene; Gibbs, 1994) and pEGFP-N1 (green fluorescent protein reporter gene; Clontech). NLS (CGGPKKKRKVG-NH2) was purchased from Eurogentec. The plasmid DNA–peptide complexes used had a molar ratio of 1:100, as described elsewhere (Collas et al. 1996).

RESULTS AND DISCUSSION

NLS mediated transgenesis

To limit the number of transgene integrations, we microinjected sub-femtogram amounts of DNA-NLS into zebrafish eggs in order to promote early nuclear uptake and germline insertion of a single copy of the transgene. Here we report single copy integrations into the zebrafish germline, inheritance and high frequency of expression of a luciferase reporter gene after cytoplasmic injection of as low as 10 copies (corresponding to 0.06 fg DNA) of a non-covalent DNA-NLS complex. Our results suggest that non-covalent binding of nuclear targeting peptides to vector DNA may prove highly valuable in non-viral transgenesis, as well as gene therapy applications.

Decreasing concentrations (10^4 to 10 copies) of pCMVL-NLS complexes were injected into the cytoplasm of newly fertilized eggs, and proportions of transgenic F_0-generation (parental) zebrafish were determined by PCR analysis of tail biopsies of 6- to 12-month-old animals using luciferase transgene-specific primers. The data suggest that a high proportion of transgenic fish can be produced with as low as 100 copies injected. Injection of 10 copies of DNA-NLS produced transgenic fish at a frequency of 20%. The use of $>10^3$ DNA-NLS complexes increases the number of transgene integrations to over 90% (Table 2.1).

Table 2.1: Efficiency of transgenesis in zebrafish following injection of pCMVL-NLS complexes [a-c]$P<0.01$ (Chi-square test) [d-f]$P<0.05$ (Chi-square test)

Group	*Transgenic F_0 (%)*	*Germline-transgenic F_0 (%)*
G10	21.0 (11/53)[a]	18.2 (2/11)[d]
$G10^2$	61.3 (27/44)[b]	22.2 (6/27)[de]
$G10^3$	93.3 (42/45)[c]	30.9 (13/42)[e]
$G10^4$	93.8 (45/48)[c]	55.5 (25/45)[f]
10^4, no NLS	0 (0/52)	—

Fluorescence in situ hybridization (FISH) analyses and transmission studies show that transgene integration into the germline and somatic tissues is mosaic, and that the extent of mosaicism is negatively correlated with the amount of DNA-NLS injected (Table 2.2). FISH was performed on nuclei isolated from whole adult transgenic F_0s using a biotinylated luciferase gene-specific probe, detected with FITC-conjugated avidin. The extent of mosaicism of transgene integration in the germline of F_0 founders was assessed by PCR screening of several hundreds of individual F_1 embryos produced by F_0 wild type (wt) breeding for the presence of the transgene. DNA copy number standards were prepared by mixing known amounts of wt genomic DNA with known amounts of pCMVL (5.5 kb). Estimates of copy number integrated were made using a DNA content of $2x10^9$ bp per haploid genome (Stuart et al. 1990) (Table 2.2).

Table 2.2: Mosaicism of transgene integration

Group	*FISH-positive nuclei (% ± SD) (F_0)*[a]	*FISH Signals/nuc. (no ± SD) (F_0)*	*PCR-positive F_1*[b] *(%)*	*pCMVL copies integrated no± SD*[c]
Wt	0 ± 0	—	—	—
G10	10.8 ± 5.8[d]	1.0 ± 0.0[h]	12.0 (13/108)[d]	1.0 ± 0[h]
$G10^2$	26.4 ±11.3[e]	1.3 ± 0.3[h]	27.3 (70/256)[e]	1.3 ± 0.9[h]
$G10^3$	48.1 ±12.1[f]	1.5 ± 0.2[h]	31.1 (71/228)[e]	11.4 ± 7.9[i]
$G10^4$	87.6 ±11.1[g]	4.4 ± 0.4[i]	47.6 (201/422)[f]	44.5 ± 14.4[j]

[a] Nuclei displaying at least one luciferase FISH signal; n = 225-1016 nuclei per group, isolated from 3-6 different whole F_0 fish
[b] F_1 obtained by transgenic F_0 x wt cross. A total of 3 -10 F_0 individuals were used in each group
[c] Estimated in semi-quantitative dot-blot assays; n = 14-18 individuals analyzed per group
[d-g] $P<0.01$ (Chi-square test)
[h-j] $P<0.01$ (*t*-test)

SV40 Ori enhancement of DNA nuclear import

As an alternative to microinjection of DNA-NLS peptide complexes, we: (i) inserted the 372 bp SV40 ORI sequence into pCMVL; and (ii) deleted the SV40 ORI region from pEGFP-N1 in order to compare the nuclear uptake of the two pairs of reporter genes. The effect of the SV40 cis acting element was measured as : (i) FISH detection of plasmid copy numbers in isolated nuclei at 4 hrs post injection (p.i.); and (ii) as transient expression at 24 hours p.i. of zebrafish zygotes. Using 10^6 copies of pEGFP-N1, we could visualize a significant difference in FISH signals (see Fig. 2.1) and with 10^4 to 10^6 copies of pCMVL, there was a clear increase in luciferase activity in 24 hr embryos after microinjection with the ORI$^+$ construct (Liang et al. 2000).

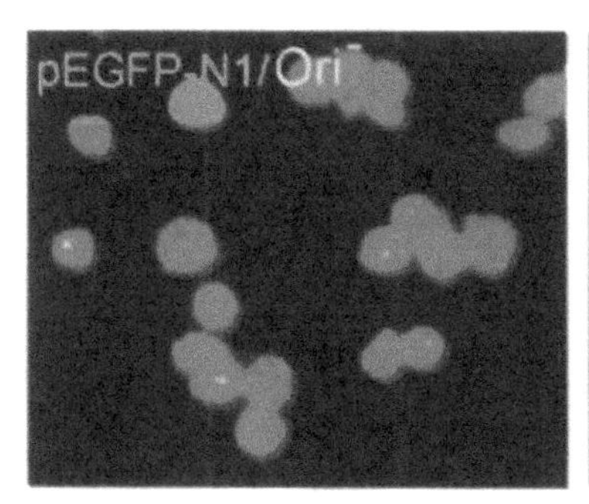

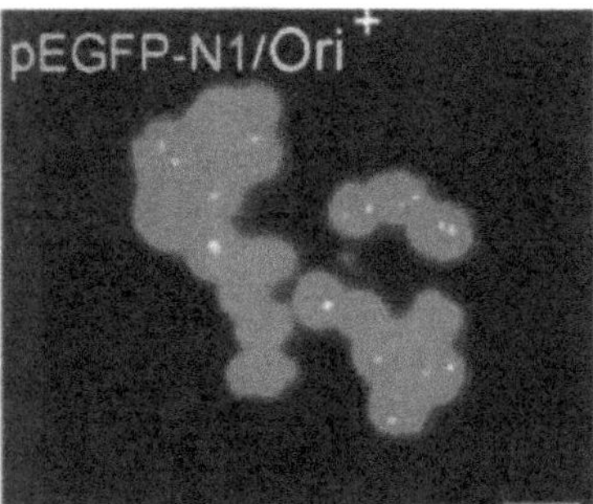

Fig. 2.1: SV40 372 bp Ori region promote nuclear import of 10^6 copies gene construct in zebrafish zygotes. Figure shows FISH signals of 4 h zebrafish embryo nuclei. Injection of 10^6 copies/egg of pEGFP-N1(Ori$^+$) compared to pEGFP-N1 with deleted ORI (Ori$^-$) dramatically enhanced the proportion of embryos expressing GFP at 24 hr as judged by fluorescence microscopy (65.8% (n = 233) and 1.4% (n = 209), respectively).

Particle-mediated gene transfer

For gene transfer to somatic tissues (e.g. dermis and epidermis) of zebrafish, conditions for gene-gun mediated biolistic transfer of luciferase (pCMVL) and green fluorescent protein (pEGFP-N1) reporter genes have been optimized. The highest overall luciferase expression was obtained in epidermis and dermis using 1 μm microcarriers coated with 1 μg pCMVL plasmid DNA, delivery at a pressure of 200 psi (Fig. 2.2). Luciferase activity within the skin peaked at 18 hours and decreased to 30% of the maximum at day 8 post DNA transfer. GFP expression was detected in epidermis already 13 min after DNA delivery and at 1 hr post-biolistic gene transfer, 100% of the cells in the target area exhibited GFP expression. One factor of importance to address when quantitative studies are to be performed using the gene gun is to solve the problem of inconsistency of DNA amount between cartridges (Torgersen et al. 2000).

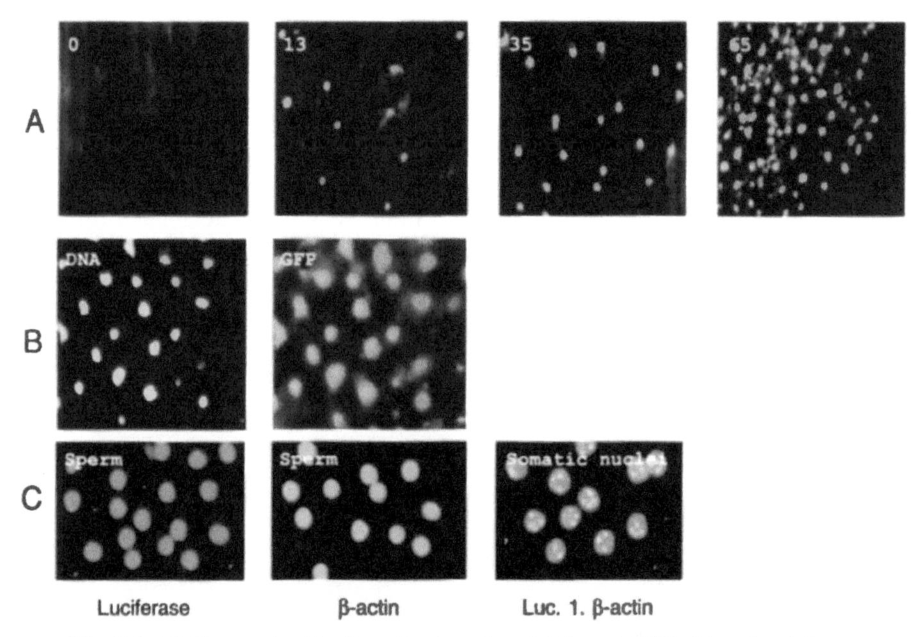

Fig. 2.2: (A) Expression kinetics after transfer of 1.0 μg pEGFP-N1, using 1.0 μm microcarriers at 200 psi. Epidermal cells were examined by fluorescence microscopy at indicated time points. (B) Epidermal cells examined for GFP expression 65 min post-DNA transfer, as described above. To determine the number of cells possessing GFP activity in the delivery target area, nuclear DNA was stained with Hoechst 33342. (C) FISH analysis using pCMVL- and β-actin-specific probes. Sperm cells were collected 15 or 30 days after gene gun-mediated transfer of pCMVL with optimal parameters. Only control β-actin gene-specific signals can be localized (red spots) in the nuclei. Control somatic nuclei isolated from pCMVL-bearing transgenic zebrafish, as described by Collas and Aleström (1998). FISH pCMVL- and β-actin-specific signals localized as green and red signals, respectively.

Accidental germline uptake

To assess the risk for accidental germline uptake, a screen of approximately 250,000 zebrafish sperm cells from 14 individuals was made by fluorescence in situ hybridization (FISH) at 15 or 30 days after delivery of 1 μg pCMVL DNA (Fig. 2.2C). The lack of one single positive transgene FISH signal argues for lack of—or extremely low frequency of—accidental transgene uptake and integration into cells of the male germline following gene gun-mediated DNA delivery (Torgersen et al. 2000).

CONCLUDING REMARK

The above expose of methods developed for gene transfer to zebrafish is not intended to be complete. The use of viral vectors and various liposome and other macromolecular complexes has not been included here, but it does represent the alternative strategies tested to achieve more efficient and better control of gene transfer to aquatic organisms. It seems obvious that future basic research will again develop novel strategies for gene transfer and future aquaculture will be dependent on such new tools for the improvement of traits controling disease resistance, food quality, etc. Some solutions will deal with somatic cell gene treatments (DNA vaccines), whereas others will probably involve germline modification. However, as important as technological development and inventions are, care and precautions with regard to health and environmental biosafety (and economy) will be while applying research results to aquaculture, to select and develop the good ideas and, at the same time, avoid the less good ones (Aleström, 1996; Holmenkollen Guidelines for Sustainable Aquaculture, 1998).

ACKNOWLEDGEMENTS

This work was supported by VESO, the European Union and the Norwegian Research Council.

REFERENCES

Aleström, P. 1996. Genetically modified fish in future aquaculture: Technical, environmental and management considerations. Turning Priorities into Feasible Programs, ISNAR-IBS, 4: 81-85.

Collas, P. and Aleström, P.1996. Nuclear localization signal of SV40 T antigen directs import of plasmid DNA into sea urchin male pronuclei in vitro. Mol. Reprod. Devel., 45: 431-438.

Collas, P. and Aleström, P. 1997. Rapid targeting of plasmid DNA to zebrafish embryo nuclei by the nuclear localization signal of SV40 T antigen. Mol. Mar. Biol. Biotech., 6: 48-58.

Collas, P. and Aleström, P. 1998. Nuclear localization signals enhance germline transmission of a transgene in zebrafish. Transgenic Res., 7:303-309.

Collas, P., Husebye, H. and Aleström, P. 1996. The nuclear localization sequence of SV40 T antigen promotes transgene uptake and expression in zebrafish embryo nuclei. Transgenic Res., 5: 451-458.

Collas, P., Husebye, H. and Aleström, P. 1997. Transfering foreign genes into zebrafish eggs by microinjection. in: Transgenic Animals—Generation And Use, (ed.) Houdebine, L.M., Harwood Academic Publ. Part II, Section D, pp 119-122.

Culp, P., Nüsslein-Vohlhard, C. and Hopkins, N. 1991. High-frequency germline transmission of plasmid DNA sequences injected into fertilized zebrafish eggs. Proc. Natl. Acad. Sci. USA, 88: 7953-7957.

Dean, D.A. 1997. Import of plasmid DNA into the nucleus is sequence specific. Exp. Cell Res., 230: 293-302.

Fritz, J.D., Herweijer, H. Zhang, G. and Wolff, J.A. 1996. Gene transfer into mammalian cells using histone-condensed plasmid DNA. Hum. Gene Ther., 7: 1395-1404.

Fynan, E.F, Robinson, H.L. and Webster R.G. 1993. Use of DNA encoding influenza hemagglutinin as an avian influenza vaccine. DNA Cell Biol., 12: 785-789.

Gibbs, P., Peek, A. and Thorgaard, G. 1994. An in vivo screen for the luciferase transgene in zebrafish. Mol. Mar. Biol. Biotech., 3: 307-316.

Gómez-Chiarri, M., Livingston, S.K., Muro-Cacho, C., Sanders, S. and Levine, R.P. 1996. Introduction of foreign genes into the tissue of live fish by direct injection and particle bombardment. Dis. Aquat. Org., 27: 5-12.

Heppell, J., Lorenzen, N., Armstrong, N.K., Wu, T., Lorenzen, E., Einer-Jensen, K., Schorr, J. and Davis, H.L. 1998. Development of DNA vaccines for fish: Vector design, intramuscular injection and antigen expression using viral *Haemorrhagic septicaemia* virus genes as model. Fish Shellfish Immunol., 8, 271-286.

Holmenkollen Guidelines for Sustainable Aquaculture 1998, Oslo, Norway November 1997. www.ntnu.no/ntva/rapport/aqua.html

Iyengar, A., Muller, F. and Maclean, N. 1996. Regulation and expression of transgenes in fish—A review. Transgenic Res., 5: 147-166.

Kaneda, Y., Iwai, K. and Uchida, T. 1989. Increased expression of DNA cointroduced with nuclear protein in adult rat liver. Science, 242: 375-378.

Liang, M-r., Aleström, P. and Collas, P. 2000. Glowing zebrafish: Single luciferase transgene integration, transmission and expression promoted by nuclear localization signals. Mol. Reprod. Dev., 55: 8-13.

Lin, S., Yang, S. and Hopkins, N. 1994. *lacZ* expression in germline transgenic zebrafish can be detected in living embryos. Dev. Biol., 161: 77-83.

Pandian, T.J. and Marian, L.A. 1994. Problems and prospects of transgenic fish production. Curr. Sci., 66: 635-639.

Pandian, T.J. and Venugopal, T. 2004. Contribution to transgenesis in Indian major carp *Labeo rohita_in*: Fish genetics and aquaculture biotechnology (eds) Pandian, T.J., Strussmann, C.A. and Marian, M.P. Oxford & IBH publishers, New Delhi. p. 1-20.

Rolland, A.P. 1998. From genes to gene medicines: Recent advances in non-viral gene delivery. Crit. Rev. Ther. Drug Carrier Syst., 15: 143-198.

Sebestyen, M.G., Ludtke, J.J., Bassik, M.C., Zhang, G., Budker, V., Lukhtanov, E.A., Hagstrom, J.E. and Wolff. J.A. 1998. DNA vector chemistry: The covalent attachment of signal peptides to plasmid DNA. Nat Biotechnol., 16: 80-85.

Stuart, G.W., McMurray, J.V. and Westerfield, M. 1988. Replication, integration and stable germline transmission of foreign sequences injected into early zebrafish embryos. Development, 103: 403-412.

Stuart, G.W., Vielkind, J.R., McMurray, J.V. and Westerfield, M. 1990. Stable lines of transgenic zebrafish exhibit reproducible patterns of transgene expression. Development, 109: 577-584.

Torgersen, J., Collas, P. and Aleström, P. 2000. Gene gun-mediated transfer of reporter genes to somatic zebrafish (*Danio rerio*) tissues. Mar. Biotechnol., 2: 293-300.

Westerfield, M. 1993. The Zebrafish Book. University of Oregon Press. The Fish Net (http://zfish.uoregon.edu/).

Wolff, J.A., Malone, W.R., Williams, P., Chong ,W., Ascadi, G., Ani, A. and Felgner, P.L. 1990. Direct gene transfer into mouse muscle in vivo. Science, 247, 1465-1468.

Zanta, M.A, Belguise-Valladier, P. and Behr, J.P. 1999. Gene delivery: A single nuclear localization signal peptide is sufficient to carry DNA to the cell nucleus. Proc. Natl. Acad. Sci. USA, 96: 91-96.

Chapter 3

Application of RAPD and AFLP to Detect Genetic Variation in Fishes

P. Jayasankar
Central Marine Fisheries Research Institute, Ernakulam North PO, Cochin-682 018.
Email: jayasankarp@vsnl.com

ABSTRACT

DNA-level markers find application in species identification, evaluation of phylogeny, delineation of stock structure, measurement of levels of genetic variations, conservation genetics, determination of breeding strategies, gene mapping and marker-assisted selection. RAPD (Random Amplified Polymorphic DNA) is a technique that reveals DNA-based arbitrarily primed polymorphisms. There is no need for prior knowledge of the target sequence. RAPD analysis is relatively simple, inexpensive and rapid, thus being preferred over the more laborious RFLP technique. RAPDs have been successfully applied in taxonomic studies, inheritance studies and for stock discrimination of marine and freshwater invertebrates and fishes. AFLP (Amplified Fragment Length Polymorphisms) technique is based on the detection of genomic restriction fragments by PCR amplification. Fingerprints are produced without prior sequence knowledge using a limited set of genetic primers. The number of fragments detected in a single reaction can be "tuned" through the selection of specific primer sets. AFLP combines the reliability of RFLPs and the power of PCR. Compared to RAPDs, AFLP markers are more in number and exhibit greater specificity and stability. AFLP markers have found application in evaluating genetic variation and linkage mapping in fishes. This chapter reviews some of the important works with these markers.

Key words: RAPD, AFLP, genetic variation

INTRODUCTION

DNA-level markers have several advantages over morphometric, meristic or protein (allozymes) markers for studying stock structure and genetic variation among fishes (Carvalho and Pitcher, 1994; Ferguson and Danzmann, 1998).

Random Amplified Polymorphic DNA (RAPD) is a DNA approach for detecting genetic polymorphisms (Welsh and McClelland, 1990; Williams et al., 1990). The technique allows the detection of DNA polymorphisms by amplifying randomly chosen regions of DNA by PCR with single arbitrary primers. Any section of DNA flanked by a pair of primer sites, and less than ~5000 base pairs apart, will be amplified by the RAPD technique. The amplified products are separated by gel electrophoresis and detected by direct staining with ethidium bromide or silver nitrate. Prior knowledge of the target sequence is not required and the technical aspect simplicity makes RAPDs attractive for population genetic studies (Jayasankar and Dharmalingam, 1997a). Further, a large number of DNA primers can be screened for population polymorphisms.

Smith (2003) has summarized RAPD studies on population genetics and molecular taxonomy of commercially important fishes and aquatic invertebrates. Numerous studies have reported the separate effects of altering different parameters, ratio of template DNA primers, concentration of *Taq* DNA polymerase and Mg concentration on the bands obtained. A corollary of these experiments is that RAPD profiles should be reproducible among laboratories provided that all details of the reaction conditions are standardized and strictly adhered to.

On the other hand, Amplified Fragment Length Polymorphisms (AFLP) (Vos et al., 1995) is a technique, with relatively higher throughput and reliability. AFLP markers have been found effective in estimating genetic variation and mapping of useful genes. Though they have been used in several plant species, their application in fishes is comparatively limited (Albertson et al., 1999; Yue et al., 2002). This chapter presents an overview of different investigations conducted using these markers in Indian fishes.

GENETIC ANALYSIS USING RAPD

Critical evaluation of RAPD protocol including DNA extraction, amplification, separation, fragment scoring and data analysis has been given elsewhere (Smith, 2003). Three major areas in which RAPD has found application in fisheries include inheritance studies, molecular taxonomy and stock discrimination.

Inheritance studies

The RAPD fragments which are not present in the source DNA can appear following amplification, due to interactions between primer and fragments during the denaturation and annealing steps of PCR (Rabouam et al., 1999). Such apparent fragment polymorphisms entail Southern blotting and hybridization to confirm fragment homology. Ideally, the allelic nature of RAPD markers should be confirmed through breeding studies, which may be impractical for many aquatic species (Smith, 2003). However, several studies in fish have shown that RAPD polymorphims conform to Mendelian expectations and are consistent with a dominant model such as with guppies (Foo et al., 1995), salmonids (Elo et al., 1997), channel catfish (Liu et al., 1998) and tilapia (Appleyard and Mather, 2000). These polymorphisms could be used for subsequent genetic linkage mapping. The linkage map can also provide numerous RAPD markers for sex determination, and also markers linked to colour patterns, disease resistance, immune response and other qualitative as well as quantitative traits, which can be used for linkage selection.

Molecular taxonomy

RAPD has shown high power of resolution in separating species complexes and sibling species, in detecting cryptic pairs of species and in confirming close relationships between species. Pooling of DNA samples from intraspecific individuals permits the rapid screening of a large number of primers for taxonomic studies and takes no more laboratory time than conventional allozyme screening (Smith, 2003).

Govindaraju and Jayasankar (2004) studied the status of *Epinephelus spp* using RAPD analysis based on samples drawn from southeast and southwest coasts of India. The RAPD fingerprints generated in *Epinephelus diacanthus, E. areolatus E. chlorostigma, E. bleekeri, E. coioides, E. tauvina,* and *E. malabaricus* with four primers (OPA 01, OPA 07, OPF 08 and OPF 10) were consistent, reproducible and yielded species-specific diagnostic markers in all the species (Table 3.1).

Table 3.1: Species-specific diagnostic RAPD markers (in bp) in Epinephelus spp. generated by four arbitrary primers (from Govindaraju and Jayasankar, 2004)

Species		*Primers*		
	OPA 01	*OPA 07*	*OPF 08*	*OPF 10*
E. areolatus	610	745	1250, 925	910, 720
E. chlorostigma	775, 670	650	820	300
E. bleekeri	660, 430	1320, 170	600	640, 120
E. coioides	875	860	815	170
E. tauvina	1610, 980	905	1130, 858	245
E. malabaricus	855	365	940	90

A total of 59 RAPD loci in the size range of 70-4500 bp were produced from all the four arbitrary primers. UPGMA dendrogram was constructed on the basis of genetic distance values to show the genetic relationships among the seven species. All the individuals of each species formed monophyletic species clusters. Mean intraspecies genetic distance value (0.305) was significantly lower than interspecies value (0.365). *E. malabaricus* was observed to be most distantly related to *E. diacanthus* and *E. bleekeri*. A very close genetic relationship was seen among *E. coioides, E. tauvina* and *E. malabaricus* and also between *E. chlorostigma* and *E. bleekeri*. Within species, genetic polymorphism was highest in *E. chlorostigma* (49.15%) and lowest in *E. tauvina* (25.42%).

In an another study (Jayasankar, 2004), species status of domesticated clown fish species from two main aquaria of Central Marine Fisheries Research Institute was ratified using RAPD. The domesticated species of the clown fish at Vizhinjam marine aquarium in the southwest coast of India was identified as *Amphiprion chrysogaster* (Gopakumar et al. 1999). However, the species, which was domesticated at Mandapam marine aquarium in the southeast coast of India, was identified as *A. sebae* (Ignatius et al., 2001). The stock of the latter was drawn from the former, thus belonging to the same species, but the species status was debated upon. RAPD analysis of individuals from both these aquaria and those of *A. chrysogaster* obtained from Mauritius collections of Gerald Allen (an expert in pomacentrid taxonomy) revealed that the Indian domesticated species was not *A. chrysogaster*. UPGMA phenerogram showed clustering of Mandapam and Vizhinjam samples together, clearly separated from Mauritius samples. All the individuals of each species formed mononphyletic species clusters. Some RAPD fragments have shown fixed frequencies, which can be used as species diagnostic markers in *A. sebae* and *A. chrysogaster*.

Stock discrimination

A basic prerequisite in fisheries management is the identification of production units or stocks of species; inadequate knowledge of stock structure may lead to over- or under-exploitation. Jayasankar et al., (2004) have applied a holistic approach, combining Truss morphometrics, protein polymorphisms and RAPD to study stock variations in Indian mackerel from peninsular India. Three RAPD primers were used in this study, which had generated a total of 59 loci varying in size from 560 to 4500 bp. None of the 3 populations from the east and west coasts of India showed RAPD fragments of fixed frequencies, to be treated as population-specific markers. Although two populations were clustered together with the third one forming an outgroup, both intra- and inter-population genetic distance values were not significant among the three populations.

For several species of fishes the stock structures, revealed with RAPDs and either allozymes or mtDNA markers, appear to be similar. In some respects, this result is surprising as the methods measure different components of the genome, which may be subjected to different evolutionary rates, and strengthens conclusions on genetic stock relationships (Smith, 2003). In contrast, greater population differentiation was found with allozymes and RAPDs than with mtDNA RFLPs in the orange roughy (Smith et al. 1997). Occasionally, RAPDs have been used when low levels of polymorphism have been found with other methods, such as allozymes (Nilsson and Schmitz, 1995).

GENETIC ANALYSIS USING AFLP

One of the most important advantages of the AFLP technique is the high number of markers that can be screened per experiment, as multiple primer combinations can yield much more markers than with RAPDs. Further, it has greater reproducibility than RAPD. The primers used to obtain AFLP markers combine two key characteristics: (i) they are complementary to the adaptor oligonucleotide, thus allowing a highly specific primer annealing; and (ii) they are selective, changing the 3′ nucleotides allows amplification of a different set of DNA fragments from the same population of pre-amplified fragments (Vos et al. 1995).

Chong et al. (2000) found that AFLP was more efficient than RAPD as a marker system for identifying genotypes within populations of Malaysian river catfish, *Mystus nemurus* due to its higher resolution power. RAPD and AFLP markers yielded comparable results of genetic diversity in Asian arowana, *Scleropages formosus* (Yue et al. 2002). AFLP was a useful marker to identify two subspecies of Ayu, *Plecoglossus altivelis* (Takagi et al., 1998) while this marker system proved useful for the assessment of intraspecific and interspecific genetic variations in *Morone* and *Thunnus spp* (Han and Ely, 2002). AFLP could generate clear species-diagnostic markers in *Thunnus* spp. The advantage of AFLP over mtDNA as a species-diagnostic marker is that the former samples the entire genome; thus it is not based on a particular allele or a particular haplotype. This point is particularly important since mtDNA-based analyses cannot be used to detect hybrids while a hybrid pattern would be obvious using AFLP.

AFLP markers were found suitable for use in gene and quantitative trait loci mapping of catfish by using the interspecific hybrid system involving *Ictalurus punctatus* and *I. furcatus* (Liu et al. 1998, 1999). The 64 primer combinations amplified 7,871 bands, of which 3,081 were polymorphic between the two mating parents of two catfish species used to produce the reference families for mapping analysis (Liu et al. 1999). AFLP was reported to be useful for verification of gynogenesis in sea bass, *Dicentrarchus labrax* (Felip et al., 2000).

Wang et al. (2000) have developed AFLP markers using 40 individuals of common carp *Cyprinus carpio* belonging to 3 different Indonesian stocks, namely Majalaya, Sinyonya and one batch of meiotic gynogens derived from Majalaya (Gynogen M). Primer combinations E-AAC/M-CAA, E-AGG/M-CAC and E-ACC/M-CTT revealed a total of 299 AFLPs in 40 fish (117, 90 and 92, respectively) with a polymorphism of 66.2% (198). Average percentage of polymorphic loci was maximum in Majalaya and minimum in Gynogen (M) E-ACC/M-CTT produced relatively high percentage of polymorphisms (Table 3.2).

Table 3.2. Genetic variability in Indonesian common carps detected by AFLP markers (from Wang et al., 2002)

Stock	*AFLP primer combinations*	*Fish (no.)*	*Genotypes (no.)*	*Bands (no.)*	*Loci polymorphism (%)*
Gynogen (M)	E-AAC/M-CAA	20	20	95	34.7 (33)
	E-AGG/M-CAC	20	19	64	29.7 (19)
	E-ACC/M-CTT	20	20	62	38.7 (24)
Majalaya	E-AAC/M-CAA	9	9	109	46.8 (51)
	E-AGG/M-CAC	10	10	86	52.3 (45)
	E-ACC/M-CTT	10	10	80	58.8 (47)
Sinyonya	E-AAC/M-CAA	10	10	107	39.3 (42)
	E-AGG/M-CAC	10	10	83	51.8 (43)
	E-ACC/M-CTT	10	10	73	45.2 (33)

As judged from these results, the most diverse group among those examined in the present work was the Majalaya carp and the least was Gynogen (M). Maximum Band Sharing Index (BSI) was exhibited by the gynogens and minimum by the Majalaya carp. Inter-stock comparisons showed the gynogens to be closest to the Majalaya and farthest from the Sinyonya stock. AFLP markers generated in common carp proved useful to test the degree of success of the induced gynogenesis and to differentiate gynogens from their parental stock.

DNA MARKER SCORING AND DATA ANALYSIS

RAPD and AFLP polymorphisms are scored by the presence (1) or absence (0) of an amplification product. Size of bands can be determined by comparison with a λ DNA digested with *Eco* RI / *Hind* III molecular weight marker. Software such as Bioprofil (Bio-1D) is used to calculate the fragment sizes of the bands with reference to molecular size markers. The 'species-specific diagnostic' markers are defined as those bands which are exclusive to a species for a given primer (RAPD) or sets of primers (AFLP).

The similarity index between all possible pair-wise comparisons of individuals is calculated using the formula:

$$S_{xy} = 2n_{xy}/(n_x + n_y)$$

where, n_x and n_y are the number of fragments in individuals x and y, and n_{xy} is the number of fragments shared between those individuals (Nei, 1978). Genetic distances between paired individuals or species (Nei, 1978) and gene diversity (Nei, 1973) are thus calculated.

Phylogenetic relationships between individuals or populations of fish species are constructed using cluster analysis. For this, the unweighted pair-group method with arithmetic (UPGMA) (Sneath and Sokal, 1973) contained in the NEIGHBOR programme of PHYLIP ver 3.57c, based on Nei's (1978) genetic distance values calculated for all primers shall be used. Data resampling (1000 replicates) and matrix calculations for bootstrap analysis are performed using WinBoot, an UPGMA-based programme (Yap and

Nelson, 1996). Bootstrap values between 75 and 95 are considered to be significant (Hillis and Bull, 1993).

REFERENCES

Albertson, R.C., Markerf, J.A. Danley, P.D. and Kocher, T.D. 1999. Phylogeny of rapidly evolving clade: The cichlid fishes of Lake Malawi, east Africa. Proc. Natl. Acad. Sci. USA, 96: 5107.

Appleyard, A.S and Mather, B.P. 2000. Investigation into the mode of inheritance of allozyme and random amplified polymorphic DNA markers in tilapia *Oreochromis mossambicus* (Peters). Aquaculture Res., 31: 435-445.

Carvalho, G.R. and Pitcher,T.J. 1994. Molecular genetics in fisheries. Rev. Fish. Biol. Fish., 4: 269-399.

Chong L.K., Tan, S.G., Siraj, S.S., Christianus, A. and Yussof, K. 2000. Mendelian inheritance of random amplified polymorphic DNA (RAPD) markers in the river catfish, *Mystus nemurus*. Malays. Appl. Biol., 28: 81.

Elo, K., Ivanoff, S., Vuorinen, J.A. and Piironen, J. 1997. Inheritance of RAPD markers and detection of interspecific hybridization with brown trout and Atlantic salmon. Aquaculture, 152: 55-65.

Felip, A., Martinez, G.R., Piferrer, F., Carrillo M. and Zanuy S. 2000. AFLP analysis confirms exclusive maternal genomic contribution of meiogynogenetic sea bass (*Dicentrarchus labrax* L.). Mar. Biotechnol., 2: 301-306.

Ferguson M.M., and Danzmann, R.G. 1998. Role of genetic markers in fisheries and aquaculture: Useful tools or stamp collecting. Can. J. Fish. Aquat. Sci., 55: 1553-1563.

Foo, C.L., Dinesh, K.R. Lim, T.N., Chan, W.K and Phang, V.P.E. 1995. Inheritance of RAPD markers in the guppy fish, *Poecilia reticulata*. Zool. Sci., 12: 535-541.

Gopakumar, G., George, R.M. and Jasmine, S. 1999. Breeding and larval rearing of the clown fish *Amphiprion chrysogaster*. Mar. Fish. Inf. Serv., T & E Ser., 161: 8-11.

Govindaraju, G.S. and Jayasankar, P. 2004. Taxonomic relationship among seven species of groupers (Genus: *Epinephelus*, Family: Serranidae) as revealed by RAPD fingerprinting. Mar. Biotechnol., 6: 229-237.

Han, K. and Ely, B. 2002. Use of AFLP analyses to assess genetic variation in *Morone* and *Thunnus* species. Mar. Biotechnol., 4: 141-145.

Hillis, D.M and Bull J.J. 1993. An empirical test of bootstrapping a method for assessing confidence in phylogenetic analysis. Syst Biol., 42: 182-192.

Ignatius, B., Rathore, G., Jagdis, I., Kandasami D. and Victor A.C.C. 2001. Spawning and larval rearing techniques for tropical clown fish *Amphiprion sebae* under captive conditions. J. Aquacul. Trop., 16: 241-249.

Jackson, K., Goldberg, D., Yehuda Y. and Degani G. 2000. Molecular variation in koi (*Cyprinus carpio*) of various color patterns. Israeli J. Aquaculture, 52: 151-158.

Jayasankar, P. 2004. Random amplified polymorphic DNA fingerprinting resolves species ambiguity of domesticated clownfish (Genus : *Amphiprion*, Family : Pomacentridae) from India. Aquaculture Res., 35.

Jayasankar, P., Thomas, P.C., Paulton, M.P. and Mathew, J., 2004. Morphometric and genetic analyses of Indian mackerel (*Rastreliger Kanagurta*) from peninsular India. Asian Fish. Sci. IT.

Jayasankar, P. and Dharmalingam, K. 1997a. Potential application of RAPD and RAHM markers in genome analysis of scombroid fishes. Curr. Sci., 72: 383-390.

Jayasankar, P. and Dharmalingam K. 1997b. Analysis of RAPD polymorphisms in *Rastrelliger kanagurta* off India. Naga, the ICLARM, 20: 52-56.

Lehmann, D., Hettwer H. and Taraschewski H. 2000. RAPD-PCR investigations of systematic relationships among four species of eels (Teleostei; Anguilludae), particularly *Anguilla anguilla* and *A. rostrata*. Mar. Biol., 137: 195-204.

Liu, Z., Nichols, A., Li P. and Dunham R.A. 1998. Inheritance and usefulness of AFLP markers in channel catfish (*Ictalurus punctatus*) and blue catfish (*I. furcatus*). Mol. Gen. Genetics, 258: 260-268.

Liu, Z., Li, P., Kucuktas, H., Nichols, A. and Tan, G. 1999. Development of amplified fragment length polymorphisms (AFLP) markers suitable for genetic linkage mapping of catfish. Trans. Am. Fish. Soc., 128: 317-327.

Nei, M. 1973. Analysis of gene diversity in subdivided populations. Proc. Natl. Acad. Sci. USA, 70: 3321-3323.

Nei, M. 1978. Estimation of average heterozygosity and genetic distance from a small number of individuals. Genetics, 89: 583-590.

Nilsson, J. and Schnitz, M. 1995. Random Amplified Polymorphic DNA (RAPD) in Arctic char. Nordic J. Fresh. Res., 71: 372-377.

Rabouam, C., Comes, A.M. Bretagnolle, V. Humbert, J-F., Periquets, G. and Bigots, Y. 1999. Features of DNA fragments obtained by random Amplified polymorphic DNA (RAPD) analysis. Mol. Ecol., 8: 493-503.

Smith, P. J., Benson, P.G. and McVeagh, S.M. 1997. A comparison of three genetic methods used for stock discrimination of orange roughy, *Hoplostethus atlanticus*: Allozymes, mitochondrial DNA, and rapid amplification of polymorphic DNA. Fish. Bull., 95: 800-811.

Smith, P.J., 2003. Genetic analysis: Random Amplified Polymorphic DNA (RAPD). In: Stock Identification Methods, Academic Press. New York, pp. 21.

Sneath, P.H.A and Sokal, R.R. 1973. *Numerical taxonomy*. W. H. Freeman, San Francisco.

Takagi, M., Sogabe, G. and Taniguchi, N. 1998. Genetic variability and divergence of ayu *Plecoglossus altivelis* using AFLP fingerprinting. Suisan Ikushu. 26: 55-61.

Vos, P., Hogers, R., Bleeker, M., Reijans, M., Van de Lee, T., Hornes, M. Frijters, A., Pot, J., Pelman, J., Kuiper, M. and Zabeau, M. 1995. AFLP: A new technique for DNA fingerprinting. Nucleic Acids Res., 23: 4407-4414.

Wang, Z., Jayasankar, P., Khoo , S.K., Nakamura, N., Sumantadinata, K., Carman, O. and Okamoto, N. 2000. AFLP fingerprinting reveals genetic variability in common carp stocks from Indonesia. Asian Fish. Sci., 13: 139-147.

Welsh, J. and McClelland, M. 1990. Fingerprinting genomes using PCR with arbitrary primers. Nucleic Acids Res., 18: 7213-7218.

Williams, G.K., Kubelik, A.R., Livar, K.L, Rafalski, J.A. and Tingey, S.V. 1990. DNA polymorphisms amplified by arbitrary primers are useful as genetic markers. Nucleic Acids Res., 18: 6531-6535.

Yap, I.V. and Nelson, R.J. 1996. WinBoot: A program for performing bootstrap analysis of binary data to determine the confidence limits of UPGMA-based dendrograms. International Rice Research Institute (IRRI), Manila.

Yue, G., Li Chen, Y., Chen, F., Cho, S., Lim, L.C. and Orban, L. 2002. Comparison of three DNA marker systems for assessing genetic diversity in Asian arowana (*Schleropages formosus*). Electrophoresis, 23: 1025-1032.

Chapter 4

Androgenesis and Conservation of Fishes*

T.J. Pandian[1] and **S. Kirankumar**[2]
[1]School of Biological Sciences, Madurai Kamaraj University, Madurai – 625 021 Email: tjpandian@eth.net
[2]Department of Embryology
Carnegie Institution of Washington, Baltimore, MD 21210 USA

ABSTRACT

Due to non-visibility of the egg nucleus, the established scheme of nuclear manipulation to clone a fish may prove to be a difficult task. However, fishes are amenable for interspecific androgenetic cloning. A recent discovery of using cadaveric sperm to successfully generate progenies has opened the possibility of adopting a simple, widely practicable method of post-mortem preserved (at -20°C) sperm to induce androgenesis. Inactivation of maternal genome by UV-irradiation and activation of genome-inactivated homologous or heterologous egg by a single diploid or 2 haploid, fresh or preserved sperm are some landmark events, which have not only accelerated research activity but also focussed the importance of androgenesis in aquaculture and conservation of fish germplasm. With the absence of acrosome in the teleostean sperm, fertilization in fish is not a species-specific event. Eggs of many teleosts are amenable for heterospecific insemination. Successful heterospecific insemination results in activation or fertilization of an ovum of an alien species and is the most important strategic step for induction of interspecific androgenetic cloning. Polyspermy, especially dispermy, occurs in nature and can be experimentally achieved after incubation of the milt in calcium chloride or polyethylene glycol (PEG). The paternal origin of androgenotes is verified using selected phenotypic, protein and/or molecular markers as well as karyotying and progeny testing. Recently, reporter genes, the green fluorescent protein gene (GFP) and the Tc1 transposan specific marker, have also been used. While confirming the paternity of androgenotes, progeny testing has also indicated the unexpected occurrence of females, which are, however, shown to carry XY genotype. Survival of androgenotes can be improved using a single diploid—rather than 2 haploid—sperm for activation. About 84% androgenotes succumb during embryonic development. Haploid genome

* Reprinted with permission from Current Science Vol. 85: 917-931 (2003)

regulates the time scale of developmental sequence in both homologous and heterologous eggs of *Puntius* spp, as effectively as that of diploid. A couple of research groups have restored a fish species using its preserved sperm along with the genome-inactivated eggs of another species. A comparison on the source, technique and genomes used for generation of clones of mammals and androgenetic clones of fishes indicates that from the point of conservation and aquaculture, interspecific androgenetic cloning has an edge over that of mammals.

Key words: Recipient egg, genome inactivation, sperm donor, heterospecific insemination, dispermy, markers, interspecific cloning.

INTRODUCTION

In biology, clones denote genetically identical progenies produced by a single parent. Clones are easily obtained in asexually reproducing simple organisms, plants and fungi. However, cloning does not occur in sexually reproducing higher organisms, whose progenies are drawn from equal genomic contributions from their two parents. Consequently, such progenies need not be necessarily genetically identical copies of either of the parents. In almost all higher animals, only the egg/zygote is totipotent, i.e. the gametes have the ability to develop into a complete individual. Briefly, the unique molecular organization of egg cytoplasm alone provides the signals to the nucleus, drawn from the egg/sperm or from a differentiated cell, to execute the programme of embryonic development. With an orderly series of divisions of the egg/zygote, the daughter cells are progressively reduced of their original totipotency to pluripotency and finally to unipotency. The molecular organization of cytoplasm of these daughter and grand daughter cells are not capable of providing appropriate signals to their respective nuclei to execute the orchestrated programme that regulates the development of a single cell into an organized multicellular entity, i.e. while these cell nucleus have the entire genetic information, their expression is controlled by signals being received from the surrounding cytoplasm (see also Lakhotia, 2002).

Cloning was actually first achieved nearly four decades ago, when experimental embryologists generated a large number of clones of a frog. The basic scheme of manipulative events, that were followed for cloning the frogs, are the same for mammals too but with the introduction of surrogate mother to solve problems arising from viviparity, and super-ovulation to increase the number of progenies at a given time. The basic scheme includes removal of haploid nucleus from the egg by microsurgery and implanting a diploid nucleus derived from a somatic cell of the animal to be cloned. The nucleus is either microinjected into an enucleated egg and allowed to fuse with the egg. Like a typical totipotent zygote, the chimeric egg develops into an individual.

In fishes, this established scheme of nuclear manipulation may prove to be a difficult task, due to the non-visibility of egg nucleus. Secondly, the nuclei drawn from somatic cells of fish appear less totipotent than those of frog or mammal, for even Lee et al., (2002)—who have enhanced the visibility of the nuclei by infecting the donor cells of zebrafish with Green Fluorescent Protein (GFP) gene—could transfer only the nuclei from blastula or embryonic fibroblasts and not from adult cells (see also Pandian, 2002). However, fishes are known for their amazing ability to tolerate genomes from haploid to heptaploid, genomic contribution from male or female parent alone and unequal contributions from the parents belonging to the same or different

Table 4.1: Landmark events in induction of androgenesis in fish.

Authors	*Achievements*	*Limitations*
Romashov and Belyaeva (1964)	First to claim androgenesis in loach	No evidence for total elimination of maternal genome
Stanley et al. (1976)	Record incidental occurrence of androgenetic grass carp, while hybridizing with common carp	No evidence for purity of androgenotes
Arai et al. (1979)	First to use γ-rays to eliminate maternal genome in salmon	No evidence for total elimination of maternal genome
Thorgaard et al. (1990)	First to show higher survival of androgenotes generated using tetraploid rainbow trout	No family lines established
Scheerer et al. (1991)	First to use cryopreserved milt to generate rainbow trout androgenotes	Observation contradictory to that of Bercsenyi et al. (1998)
Bongers et al. (1994)	First to use UV-irradiation to eliminate maternal genome in common carp	Other than colour, no marker used to confirm inactivation of maternal genome
Arai et al. (1995)	Generated androgenetic loach using natural tetraploid	No family lines established
Corley-Smith et al. (1996)	First to generate fertile androgenetic male zebrafish; confirmed its purity by RAPD, SSR, and MHC analyses	No information on clonal XX androgenetic siblings
Bercsenyi et al. (1998)	First to generate interspecific androgenetic goldfish	No information on maturity and reproduction of the androgenote
Nam et al. (2002)	First to generate transgenic androgenetic mud loach	No family lines established
Kirankumar and Pandian (2004a)	First to use cadaveric sperm to generate interspecific androgenetic rosy barb	—
Araki et al. (1995)	First to generate dispermic intraspecific androgenote of rainbow trout	0.1% survival; no information on reproduction
Kirankumar and Pandian (2004b)	First to generate dispermic interspecific androgenetic rosy barb	—

species (Pandian and Koteeswaran, 1998). The ease with which gynogenetic clones can be generated has virtually resulted in "down-pour" of publications, which have been reviewed by Cherfas (1981) and Chourrout (1982). Based on our publications (Pandian and Koteeswaran, 1998, Kirankumar and Pandian, 2003, 2004a, b, Kirankumar et al., 2003), this presentation comprehensively reviews the available literature on androgenesis in fishes. Androgenesis is a developmental process, facilitating the inheritance of exclusively paternal genome. It obligately involves 2 or 3 steps: (i) elimination or inactivation of egg's genome; (ii) dispermic (Kirankumar and Pandian, 2004b) or monospermic activation of embryonic development by haploid (Scheerer et al., 1986) or diploid (Thorgaard et al., 1990) gamete; and/or (iii) restoration of diploidy by suppression of the first mitotic cleavage, when embryonic development is activated by a haploid sperm (Fig. 4.1). Table 4.1 summarizes the landmark events in induction of androgenesis in fishes.

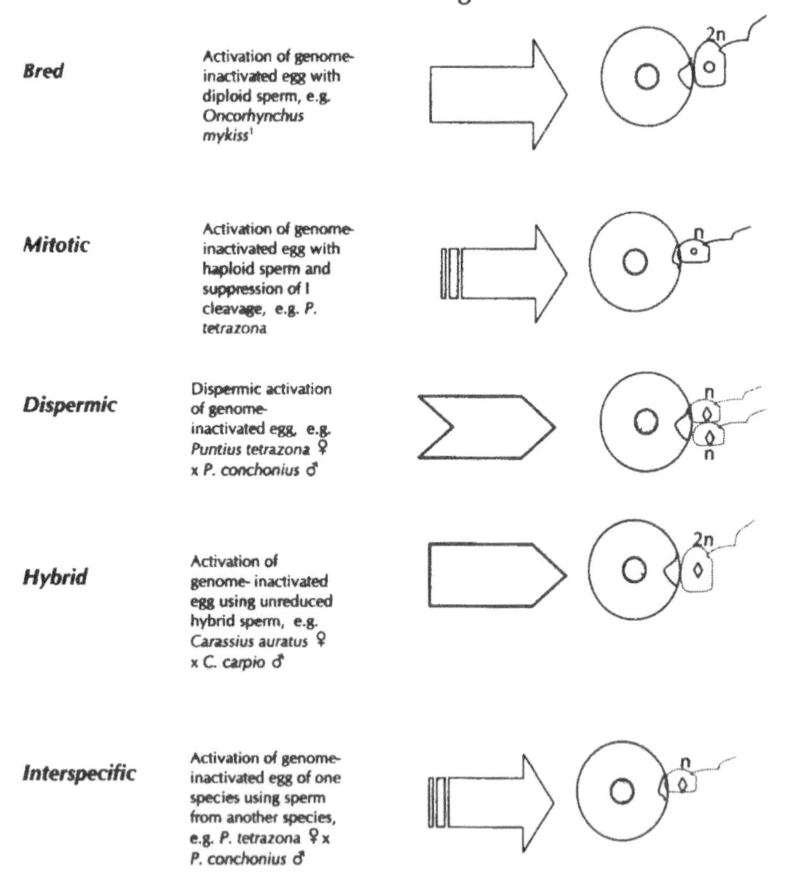

Fig. 4.1: Established protocols for induction of androgenesis in fishes. - normal activation - activation followed by thermal shock; - dispermic activation; - activation by unreduced diploid sperm

Androgenesis may prove useful for the production of: (i) viable Y^2Y^2 supermale in male-heterogametic species and Z^1Z^1 superfemale in female-heterogametic species; (ii) inbred isogenic lines; and (iii) intraspecific and interspecific androgenetic clones for conservation germplasm. Of course, YY male is known not to survive in some species (e.g. *Betta splendens*; George et al., 1994). However, androgenetic clones have been successfully generated and viable Y^2Y^2 clones have been obtained in a few species of cyprinids, cichlids and salmonids (Table 4.2). Considering the immense potential for the production of new strains in commercially important fish and loss of available strains/species due to anthropogenic activity, there is a need to develop new techniques to conserve them.

GAMETE PREPARATION

Recipient Egg

As already indicated, the non-visibility of egg nucleus of many fishes renders them non-amenable for enucleation and elimination of maternal genome of the egg. Consequently, the induction of androgenesis obligately involves irradiation of the fish egg (Table 4.2). Initially, the irradiation was limited to the use of γ rays at the dose ranging from 36 (Scheerer et al., 1986) to 88 Kr (May et al., 1988); Russian scientists like Grunina et al. (1991, 1995) used X-ray at the doses 25-30 Kr (Grunina et al., 1995). However, the irradiation may completely (Parsons and Thorgaard, 1985) or partially (Chourrout and Quillet, 1982) destroy the chromosomes. Carter et al., (1991) doubted the total elimination of egg's genome, since mtDNA and mRNA are present in large quantities in the egg (Gardener et al., 1991). Owing to protection by mitochondrial membrane, the mtDNA of the eggs of *Oreochromis niloticus* did not suffer any damage from UV-irradiation (Myers et al., 1995). Consequently, the treated eggs may still transfer some genetic material to F_1 progenies. This sort of accidental transfer of chromosomal fragments results in the undesired "genomic impurity" of the androgenotes (Arai et al., 1995). Because of their high penetrance, the irradiation by γ or X-ray is also shown to destroy 'the maternal products' like proteins (e.g. enzymes), RNA (mainly mRNA) and mtDNA (Bongers et al., 1995; Masaoka et al., 1995), obligately required for earlier development (Davidson, 1986). For instance, Stroband et al. (1992) demonstrated that these maternal products control the development in common carp zygotes until the stage of epiboly, which occurs 5-6 hrs after fertilization. Despite the cost and skill required for X-ray and γ irradiation, one of them was used to eliminate the maternal genome of the eggs, especially in salmonids, until 1991.

Bongers et al., (1994) were the first to claim 100% inactivation of the genome from *Cyprinus carpio* eggs by UV irradiation alone (Table 4.1). The carp eggs were immersed in synthetic ovarian fluid and exposed to UV radiation at the dose of 250 mJ/cm². Manual rotation of the eggs to expose

Table 4.2: Protocols used to eliminate female genome and restore diploidy for induction of androgenesis in fishes (from Pandian and Koteeswaran, 1998; modified and added)

Species	*Inactivation of female genome*	*Genetic marker*	*Survival (%)*	*Sperm source/ remarks*
		Salmonidae		
Oncorhynchus mykiss	^{60}Co; 36 kR	Isozymes	hatching : 7 feeding : 5	Inbred
			hatching : 9 feeding : 7	Outbred
O. mykiss	^{60}Co; 40 kR	—	hatching : 1 feeding : <1	Diploid
	^{60}Co; 40 kR	—	hatching : 12 feeding : 10	Tetraploid
O. mykiss	^{60}Co; 36 kR	Isozymes, Colour	hatching : 1.3 feeding : 1.0	Cryopreserved
Salvelinus fontinalis	^{60}Co; 88 kR	Allozymes	? : 38	—
		Cyprinidae		
Cyprinus carpio	X-ray; 25-30kR	Colour	hatching : 9	Inadequate genome elimination
C. carpio	UV; 100-250mJ/cm^2	Colour	hatching : 15 24 days : 10	Irradiation of eggs in ovarian fluid
C. carpio	X-ray	—	hatching : ?	*C. auratus gibelio* sperm
C. carpio	UV-?	—	—	gold fish sperm; hybrid eggs
Danio rerio	X-ray; 10000 R	RAPD, SSR, MHC	24 hr after fertilization: 22	*Danio rerio*
C. carpio	^{60}Co γ rays; 25 kR	Colour, barbel, tail morphology	hatching : 37.2	Goldfish sperm
Puntius tetrozona	UV	Fin morphology, PCR analysis, colour	hatching : 15 maturity : 7	*P. conchonius*; revived from preserved sperm (-20°C)
P. conchonius	UV	Colour	hatching : 7	100% elimination of egg genome

		Cobitidae		
Misgurnus anguillicaudatus	UV; 7500 ergs/mm^2	Allozyme	hatching : 8	100% elimination of egg genome; 2n sperm
		Cichlidae		
Oreochromis niloticus	UV	Colour	hatching : 3	fresh or cryopre-served sperm
		Characidae		
Hemigrammus caudovittatus	UV	Colour	hatching : 7	-
		Others		
Acipenser ruthenus	UV	—	hatching : ?	sperm of *A. baeri*

the animal pole to radiation yielded better results than mechanical rotation. To focus the irradiation on animal pole, Arai et al., (1995) exposed *Misgurnus anguillicaudatus* eggs to UV-source from top and bottom. Owing to the pear-like shape of eggs of the bitterlings *Rhodeus ocellatus ocellatus*, the animal pole of egg is always oriented upwards ensuring complete inactivation of egg's genome, even when UV-irradiation source is limited to the top only (Idomoto and Ueno, 1987).

UV irradiation causes several types of damage, including pyrimidine-dimers, DNA-DNA cross-links, pyrimidine adducts and others in many species. However, pyrimidine-dimer formation (T-T, C-T, C-C) in adjacent DNA-bases is the most common type of UV-damage (Friedberg, 1985). Similar damage occurs in RNA regarding the pyrimidine's uracil and cytosin. In teleosts, the pyrimidine-dimer formation is repaired by the enzyme DNA-photolyase under the influence of visible light, specifically at 300-600 nm. This enzyme also repairs damage in RNA (Friedberg, 1985). To prevent photo-reactivation of the inactivated chromosomes of gametes, the entire procedure of irradiation and diploidization is usually completed under total darkness. Myers et al., (1995) made Southern analyses of mtDNA from control and UV-irradiated eggs of *O. niloticus* in order to assess the extent of damage, and found no difference between their autoradiograms. However, the positive controls (purified mtDNA irradiated directly with a 254-nm lamp) revealed extensive damage to the mtDNA. Due to the relative position of the egg pronucleus and the scattered distribution of mitochondria throughout the egg cytoplasm, the pronucleus perhaps suffers greater damage and even total inactivation, while a large number of mitochondria remains intact either partially or totally.

Hitherto used intensity of irradiation ranges from 100-7500 ergs/m^2 (Masaoka et al., 1995). Unfortunately, many authors have not even indicated the intensity (e.g Myers et al., 1995) and the duration of UV irradiation. In general, the eggs of *Puntius tetrazona* (Kirankumar and Pandian, 2003), *P. conchonius* (Kirankumar and Pandian, 2004a), *Hemigrammus caudovittatus* and

Gymnocorymbus ternetzi (David and Pandian, 2004) are spherical and measure 0.8-1.5 mm in diameter. An intensity of 4.2 W/m^2 and a duration of 3.5 min. were found adequate to inactivate the maternal genome in the eggs of these species. However, it is likely that the optimal duration required for inactivation of maternal genome may vary from species to species, depending upon the size and cytoplasmic content of egg as well as the egg shape and position of animal pole, when the eggs are arranged in a single layer.

Ever since Bongers et al. (1994) demonstrated the effective inactivation of maternal genome by UV irradiation, it has been the choice for inactivation of the maternal genome in fishes belonging to Cyprinidae, Cobitidae, and Characidae. However, the adequacy and effectivity of the UV irradiation remain to be tested in many other groups of teleosts (Table 4.2). When a donor sperm of the same species/strain is used to activate the determination of optimal dose and duration of UV irradiation, at which 100% haploids are generated from the genome-inactivated eggs, is the procedure used to confirm the inactivation of maternal genome. The determination of haploidy requires karyotyping even at the embryonic stage, as these haploids/aneuploids suffer heavy mortality owing to haploid syndrome. For instance, following this procedure, Kirankumar and Pandian (2003) determined that at the intensity of 4.2 W/m^2, UV irradiation for 3.5 min. duration was optimal for inactivation of genome of *P. tetrazona* egg. Essentially, the procedure may prove tedious, but it is still followed by a number of scientists (e.g. Scheerer et al., 1991). However, the most commonly used procedure is to select the donor sperm of a different strain/species characterized by recessive colour and to determine the dose and duration of UV irradiation required to inactivate the maternal genome of eggs, as evidenced by the recessive coloured progenies in a cross, in which the female was characterized by the dominant colour (Table 4.3).

Table 4.3: Colour as marker in induction of androgenesis. Capital alphabet within bracket indicates the dominant (D) and recessive (R) colours (from Pandian and Koteeswaran, 1998; modified and added)

Species	*Female*	*Male*	*Progeny*
Puntius conchonius	Gray (D)	Gold (R)	Gold
P. tetrazona	Gray (D)	Blond (R)	Blond
Cyprinus carpio	Black (D)	Blond (R)	Blond
C. carpio	Black (D)	Orange (R)	Orange
C. carpio	Normal (D)	Yellow (R)	Mostly yellow
Misgurnus anguillicaudatus	Black (D)	Orange (R)	Mostly orange; few black fry (0-10.6%)
Oncorhynchus mykiss	Black (D)	Albino (R)	Spectrum of colour
Hemigrammus caudovittatus	Black (D)	Albino (R)	Albino

Sperm as genome donor

In fishes, the milt can be obtained by stripping. However, many silurids are known to be non-amenable for stripping (Na-Nakorn, 1983). Hence, milting these silurids has to be obligately invasive. Cryopreservation of fish sperm is possible (Stoss, 1983) and a large number of publications are available on choice of extenders (e.g. Scott and Baynes, 1980), cryoprotectants (e.g. Palmer et al., 1993) and other parameters for long-term sperm preservation of one or other species of fish—Cichlidae (Chao et al., 1987); Salmonidae (Lahnsteiner et al., 1996); Cyprinidae (Lubzens et al. 1997).

Sperm Preservation

Scheerer et al., (1991) are perhaps the first to show that the cryopreserved sperm can be used to induce androgenesis (Table 4.1). An important objective of the androgenesis is to use the technique for conservation of fish genome by preserving the milt of the desired species/strain, and restoring it by using genome-inactivated eggs of a suitable fish species. A couple of available publications on cryopreservation of sperm for induction of androgenesis report fragmentary and contradictory observation. Scheerer et al., (1991) indicated 1.3 and 3.8% survival of the androgenetic clones of *Oncorhynchus mykiss* generated using cryopreserved and fresh sperm, respectively. Conversely, Bercsenyi et al., (1998) indicated that the survival of the interspecific androgenetic clones of *Carassius auratus* was higher (23%) for those generated using cryopreserved sperm than that (19%) using fresh sperm. A reason for paucity of information in this area may be traced to the non-availability of liquid nitrogen facility, even in fairly big cities of developing countries. In fact, the need for expensive equipment, including the liquid nitrogen facility, has been the bottleneck in cryopreservation of fish sperm. For instance, India has a wealth of 2,118 fish species and a large number of strains in 50 commercially important fish species. However, for want of a simpler and more widely practicable protocol for preservation of fish sperm, the National Bureau of Fish Genetic Resources, Lucknow, has a sperm bank facility to hold only about a dozen fish species: *Catla catla, Labeo rohita, L. dussumieri, Cirrhina mrigala, C. carpio, O. mykiss, Salmo trutta, Tor putitora, T. khudree, Tenualosa ilisha* and *Harabagus brachysoma*. Using the preserved sperm, 65-100% hatching success has been achieved (see Ayyappan et al. 2001).

At Madurai Kamaraj University, progenies of the Indian catfish *Heteropneustes fossilis* were generated using live, fertile sperm drawn from specimens that were post-mortem preserved at –20°C for longer than 240 days (Koteeswaran and Pandian, 2002). A taxonomic survey indicated the successful use of 'cadaveric sperm' to fertilize or activate the development in eggs of many freshwater and marine fish. However, it was not clear whether these cadaveric sperm induced gynogenesis rather than syngamy of pronuclei of egg and sperm. Hence, Kirankumar and Pandian (2004a) made a comparative study using fresh and cadaveric sperm drawn from 2 different strains of *P. conchonius* characterized by a combination of 2 recessive traits, namely golden colour and unspotted tail and the other with the dominant

Table 4.4 : Sex distribution among F_1 progenies of the rosy barb generated from rosy barb eggs activated by sperm from: (a) randomly selected normal males; or (b) 30-day post-mortem preserved specimens of the rosy barb. Bold letters indicate the mean values (from Kirankumar and Pandian, 2004a).

Sperm source (strain/no.)	*Identification of the dams used for crossing (no.)*	*Hatchlings (no.)*	*Sex distribution (no.)* ♀ (X^1X^2)	♂ (X^1Y^2)
(a) Normal sires	4	83	41	42
Strain 4 ♂ 3	9	79	38	41
	3	76	38	38
		Mean	**39**	**40**
	1	88	45	43
Strain 2 ♂ 7	2	81	41	40
	4	86	38	48
		Mean	**41**	**44**
(b) Post-mortem preservation	9	18	9	9
Strain 4 ♂ 2	6	16	10	6
	5	17	8	9
		Mean	**9**	**8**
	8	19	11	8
Strain 2 ♂ 14	10	17	9	8
	1	21	12	9
		Mean	**11**	**8**

gray colour and the spotted tail. Although post-mortem preserved cadaveric sperm suffered significant losses in count, motility and fertilizability (cf. Koteeswaran and Pandian, 2002), the milt of *P. conchonius* still ensured fertilization and hatchability, which were lower (16-21%) than those generated (76-88%) using fresh sperm. Irrespective of whether fresh or cadaveric sperm were used, the sex ratio of the progenies of normal sires, and sires resulting from cadaveric sperm remained 1♂: 1♀ (Table 4.4). The sex ratio of the F_1 progenies confirmed no selective damage or mortality to X or Y carrying sperm. Hence, the discovery of post-mortem preservation of sperm at -20°C has opened the possibility of using a simple, widely practicable method of sperm preservation. This technique has special implication in India, where liquid nitrogen facility is not available in every town, especially in the north-east and western ghats, where the need for sperm preservation of rare endemic fishes is urgently required. Also, a study is yet to be undertaken to compare the fertilizability of sperm drawn from fresh milt, cryopreserved milt and post-mortem preserved (-20°C) specimens for different durations.

Heterospecific insemination

Typically, the spermatozoa of teleostean fishes do not have the acrosome

(Afzelius, 1978). However, the absence of acrosome coincides with the presence of micropyle in the eggs (Ginsburg, 1972). Since the entry of the sperm is made possible through micropyle during fertilization (Fig. 4.2), fertilization in fishes is not a species-specific event. For instance, tilapia eggs can be activated by carp (Cyprinidae) sperm (Varadaraj, 1990); eggs of *Betta splendens* (Anabantidae) can be activated by milt of *O. mossambicus* (Cichlidae; Kavumpurath and Pandian, 1994). Thus, eggs of a number of species are amenable for heterospecific insemination. Successful heterospecific insemination results in activation or fertilization of an ovum of an alien species and is the most important strategic step for induction of hybridization, hybridogenesis, gynogenesis, androgenesis and interspecific cloning, which may be defined as below: (i) hybridization, where heterospecific insemination results in the fertilization of ovum of a fish species by sperm of another fish species and production of viable hybrid progenies; (ii) hybridogenesis, in which heterospecific insemination results in hybridization but with almost total elimination of paternal chromosomes of the previous generation; (iii) gynogenesis, in which heterospecific insemination results in activation of development in haploid or diploid ovum; and (iv) interspecific androgenesis, in which heterospecific insemination results in activation of the genome-inactivated ovum.

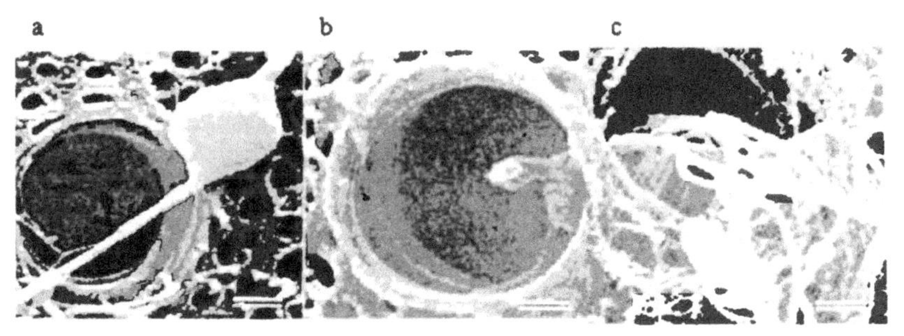

Fig 4.2: Scanning electron micrograph of: (a) unfertilized egg (30 sec. after contact) with sperm suspension in seawater; (b) fertilized egg (20 min. after contact) with sperm suspension (note the sperm tail in micropyle); and (c) fertilized egg (note the presence of many sperm in the micropylar canal) of the Atlantic herring *Clupia harengus* (from Rosenthal and Odense, 1989)

Table 4.5 presents certain selected examples of fishes, in which heterospecific insemination results in the production of viable progenies. Hitherto, experiments on heterospecific insemination have mostly been limited to commercially important food and ornamental fishes. From available information, *C. carpio* is a universal donor, whose sperm is accepted by a dozen species belonging to Cyprinidae, Cichlidae and others. Among salmonids, *O. mykiss* is perhaps a universal recipient. Reciprocal heterospecific inseminations have been a success among the following pairs: *C. carpio* and *Ctenopharyngodon idella*, *C. carpio* and *Hypophthalmichthys molitrix*. Not surprisingly, hybridization among the Indian major carps is prevalent and has led to genetic retrogression (Ayyappan et al., 2001). Such reciprocal heterospecific insemination does not necessarily result in hybridization but may induce gynogenesis, as in the case of *Puntius conchonius* ♀ x *P. tetrazona* ♂, or may also induce paternal (*C. idella* ♀ x *Hypophthalmichthys nobilis* ♂, Cassani et al., 1984) or maternal triploidy, as in many cyprinids and salmonids. *Pangasius sutchi* successfully donates sperm to *Clarias macrocephalus* but does not serve as a recipient to the sperm of *C. macrocephalus* (Na-Nakorn et al., 1993). Likewise, *P. conchonius* can be a sperm donor to *P. tetrazona*, whose sperm is, however, not acceptable to *P. conchonius* (Pandian and Kirankumar, 2003).

Polyspermy

Fish eggs are also amenable to polyspermy. Mantelman (1969) observed that 5% of just-fertilized eggs of *C. idella* contained 3 or more pronuclei and centromeres. When female *Fundulus heteroclitus* was crossed with male *Menidia notata*, more than 50% of the hybrid eggs were dispermic (Moenkhaus, 1904). Similar records on induced dispermy in eggs of triploid rainbow trout (Ueda et al., 1986) and triploid carp (Wu et al., 1993) were published. Chemicals like polyethylene glycol (PEG) and calcium chloride (see Araki et al., 1995) are known to facilitate the entry of two or more sperm into an egg. Grunina et al., (1995) explored the possibility of dispermic activation to reduce the homozygosity and generation of diploid androgenotes. They claim to have achieved the desired dispermic (*C. carpio*) activation of genome-inactivated eggs of the hybrid *C. carpio* ♀ × *C. auratus* ♂. The objective of dispermic activation of genome-inactivated eggs to generate androgenetic clones is to improve hatchability and survival. However, survival of such dispermic androgenotes is too low (e.g. 0.1%, Araki et al., 1995). Figure 4.3 presents a protocol adopted by Kirankumar and Pandian (2004b) for successful dispermic activation of genome-inactivated eggs of *Puntius* sp.

Markers

Although androgenetic clones of fish are claimed to occur in nature (Stanley et al., 1976) and artificially generated (e.g. Ponniah et al., 2001), they have failed to confirm the integrity of androgenotes. The paternal origin of androgenotes is usually verified by inspecting the progenies for selected phenotypic, protein and/or molecular markers as well as karyotyping and progeny testing. By and large, most investigators have stuck to phenotypic markers such as colour (see Table 4.3). The design is to choose a recessive colour for the male and a dominant one for the female, so that even a tint of colour present in the progeny can easily be visually detected. The total absence and presence of even the tint of the recessive colour in the progeny may

Table 4.5: Heterospecific insemination in fishes from Pandian and Koteeswaran (1988); modified and added

Sperm donor	*Sperm recipient*
Cyprinus carpio	*Ctenopharyngodon idella*
	Carassius auratus
	Hypophthalmichthys molitrix
	Cirrhinus mrigala
	Misgurnus anguillicaudatus
	Cobitis biwae
	Tinca tinca
	Oreochromis niloticus
	O. mossambicus
C. auratus	***C. idella***
	M. anguillicaudatus
	O. niloticus
C. idella	*C. carpio*
H. nobilis	***C. idella***
Puntius conchonius	***P. tetrazona***
P. gonionotus	***C. carpio***
C. biwae	***M. anguillicaudatus***
M. anguillicaudatus	***C. biwae***
M. mizolepis	***Paralichthys olivaceus***
Barbus barbus	***C. carpio***
T. tinca	***O. niloticus***
Pangasius schwanenfeldii	***P. gonionotus***
P. sutchi	***Clarius macrocephalus***
Ictalurus furcatus	***I. punctatus***
Gnathopogan elongatus elongatus	***M. anguillicaudatus***
Herichthys cyanoguttatus	***O. mossambicus***
Osteochilus hosselti	***C. carpio***
Acanthopagrus schlegeli	***P. olivaceus***
Pagrus major	***Sparus aurata***
Menida	***Fundulus***
Acipenser ruthensis	***Huso huso***
A. baeri	***A. ruthensis***
Salmo salar , S. trutta, Salvelinus fontinalis O. kisutsch, O. tshwystcha, O. masou	***Oncorhynchus mykiss***
S. trutta	***S. salar***
Thymallus thymallus	***O. mykiss***
Abramis brami	***C. carpio***
S. fontinalis	***S. trutta***
Poecilia velifera	***P. sphenops***
P. sphenops	***P. velifera***
Oreochromis aureus	***O. niloticus***
O. hornorum	***O. niloticus***
O. hornorum	***O. mossambicus***
O. macrochir	***O. niloticus***
O. variabilis	***O. niloticus***
O. hornorum	***O. aureus***
O. vulcani	***O. aureus***
O. niloticus	***O. leucostictus***
O. niloticus	***O. spilurus niger***
O. niloticus	***O. mossambicus***
O. aureus hornorum*	***O. niloticus***

Contd...

O. mossambicus	***O. spilurus niger***
Prinotus paralatus	***P. alatus***
Semotilus atromaculatus	***Phoximus oreas***
Gila eremica	***G. ditaenia***
Micropterus dolomieui	***M. salmoides***

* - hybrid resulting from the cross between *O. aureus* and *O. hornorum*

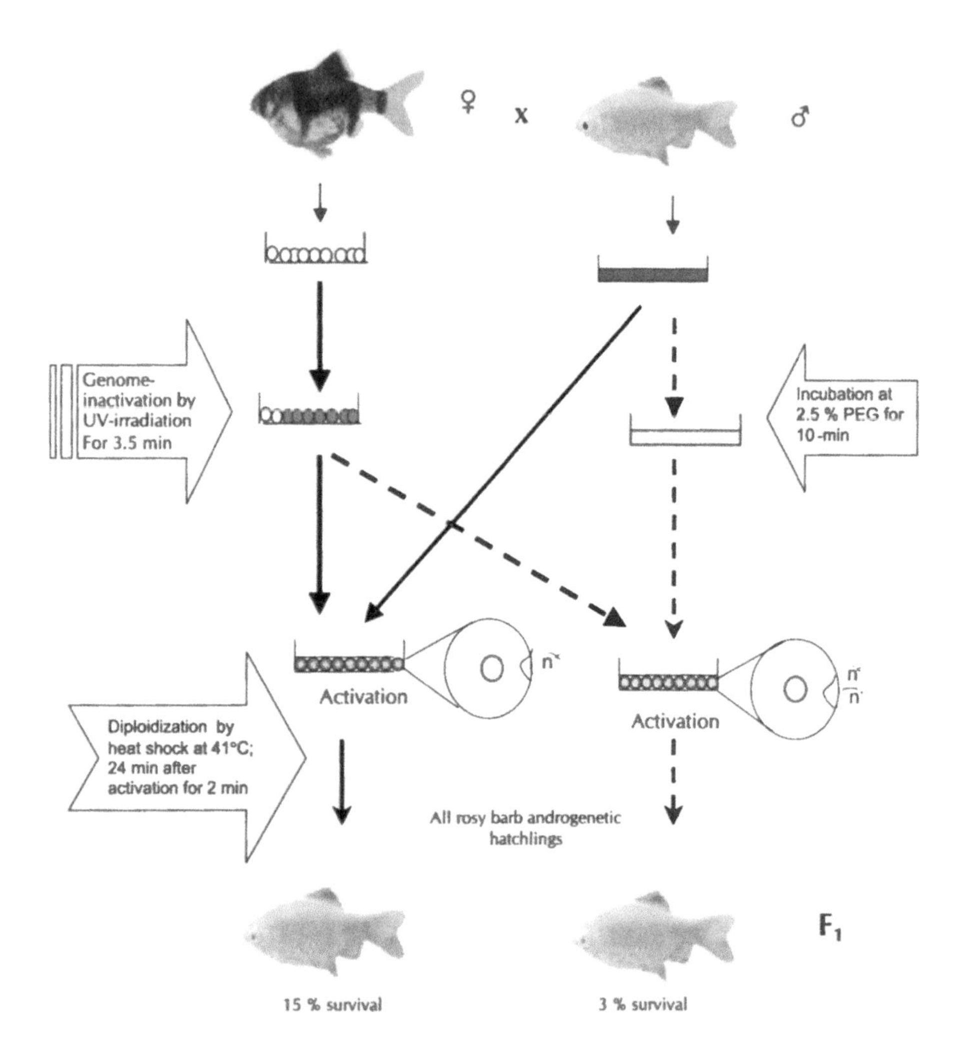

Fig. 4.3: Dispermic activation of the genome-inactivated eggs of the grey tiger barb Puntius *tetrazona* for interspecific androgenetic cloning of the golden rosy barb *P. conchonius* using its sperm incubated at 2.5% PEG for 10 min. (from Kirankumar and Pandian, 2004b)

indicate the elimination/inactivation of maternal genome and the occurrence of the undesired fractions of maternal genome, respectively. For instance, Disney et al., (1987) observed that a spectrum of colour is inherited by the F_1 progeny of *O. mykiss*, clearly indicating the incomplete elimination of maternal genome. Very few researchers like Bercsenyi et al., (1998) and Kirankumar and Pandian, (2004a) have chosen more than one phenotypic markers to confirm the paternity of the respective androgenotes (Table 4.6). A few others have also chosen protein markers like the isozyme (Scheerer et al., 1986) or allozyme (May et al., 1988; Arai et al., 1995; Scheerer et al., 1991) have chosen both colour and isozymes to confirm the paternity of *O. mykiss* androgenotes.

More recently, molecular markers have been used to irrefutably confirm the paternity of androgenotes. Kirankumar and Pandian (2003) were perhaps the first to use the Green Fluorescent Protein (GFP) gene—a reporter gene as a marker to confirm the exclusive paternal origin of the haploid. It has a couple of advantages: (i) the destiny and distribution of the paternal genome can be traced from early embryonic stages, as early as 16-hr-old embryo in *P. tetrazona* (Fig. 4.4); and (ii) it can be used to confirm the paternal origin of haploid androgenotes, even those which have succumbed at the embryonic stage.

Table 4.6: Homozygous dominant and recessive traits used as phenotypic markers to confirm the paternity of androgenotes of *Puntius conchonius* strains (from Kirankumar and Pandian, 2004a) and *Carassius auratus* strains (Bercsenyi et al. 1998)

	Puntius conchonius			
Phenotypic marker	*Strain 1*	*Strain 2*	*Strain 3*	*Strain 4*
Tail morphology	Veil tail - Dominant	Normal tail - Recessive	Normal tail - Recessive	Normal tail – Recessive
Body colour	Gray – Dominant	Gray – Dominant	Gold – Recessive	Gold – Recessive
Spot near *caudal peduncle*	Absent – Recessive	Present – Dominant	Present – Dominant	Absent – Recessive
	Carassius auratus			
	Strain 1	Strain 2	Strain 3	
Head morphology	Red cap - Recessive	Bubble eyes – Recessive	Telescope eyes – Recessive	
Body colour	White – Recessive	Red – Recessive	Black – Recessive	

Whereas phenotypic markers, isozymes and allozymes are limited to one or few genes and/or alleles, RAPD analysis provides a more comprehensive picture of the genome. Therefore, it may be a good idea to go for RAPD analysis (Corley-Smith et al., 1996), besides one or more phenotypic markers

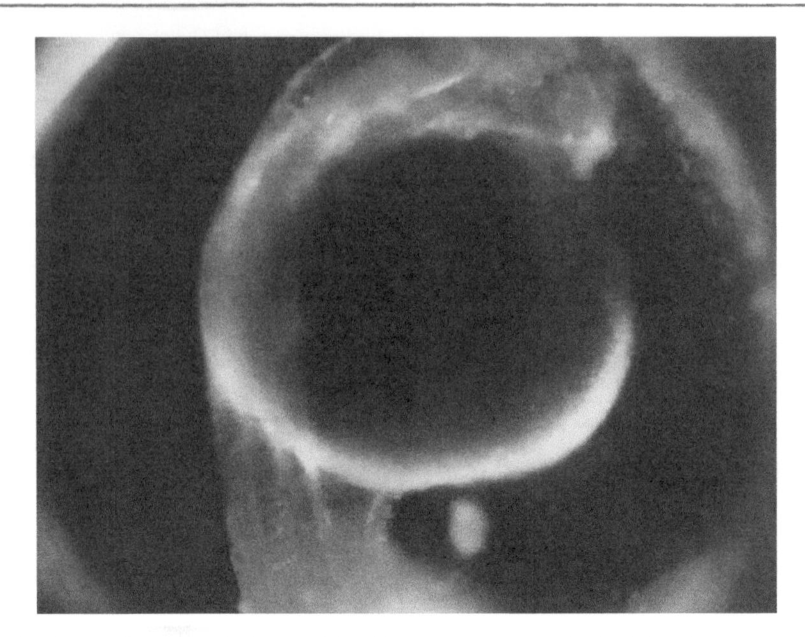

Fig. 4.4: EGFP expression in the 16-hr-old haploid androgenetic blond *Puntius tetrazona* embryo (from Kirankumar and Pandian, 2003)

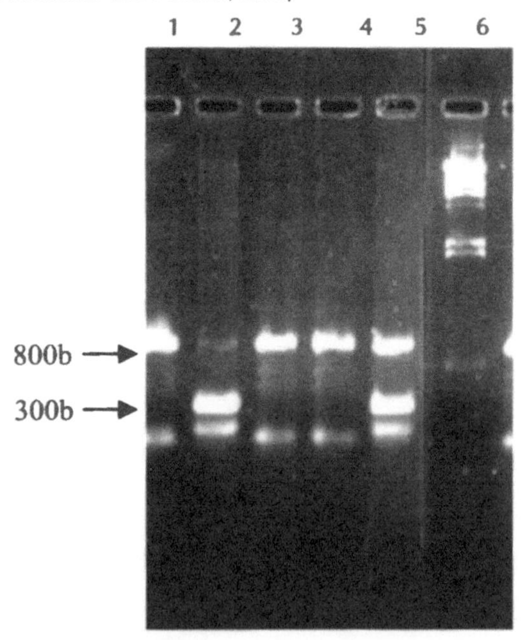

Fig. 4.5: Agarose gel electrophorogram showing species-specific Tc1 amplifications (size variants) in the rosy barb, *P. conchonius*, the tiger barb, *P. tetrazona*, the hybrid barb and the androgenetic clones. Lane1 - λ *Hind* III marker; Lane 2 – Rosy barb; Lane 3 – Tiger barb; Lanes 4 & 5 – Androgenetic clones; Lane 6 – Hybrid barb (from Kirankumar and Pandian, 2004a)

(e.g. Berscenyi et al. 1998). Kirankumar and Pandian (2004 a,b) generated monospermic and dispermic androgenetic clones of the rosy barb using genome-inactivated eggs of the tiger barb. For the first time in fishes, Tcl-transposon specific primers were used to confirm the total inactivation of maternal genome, especially when interspecific androgenotes are induced. Transposons are mobile DNA elements that are widespread components of the genomes of most organisms. Tcl-like transposons belonging to class II occur widely in the genome of fishes (Goodier and Davidson, 1994). When PCR analyses were made using Tcl transposon specific primers, the rosy barb genomic DNA produced an intense 800 bp product and the tiger barb genomic DNA an intense 300 bp product. The hybrids between these two barbs (tiger barb ♀ x rosy barb ♂) produced both these products. Expectedly, the genomic DNA of the interspecific androgenetic clones of the rosy barb, *P. conchonius* resulting from monospermic or dipsermic activation of the genome-inactivated eggs of the tiger barb, produced an intense 800 bp product only, confirming the expected paternal inheritance (Fig. 4.5). Therefore, this PCR analysis confirmed the purity of paternal genome inheritance by *P. conchonius* through the surrogate eggs of *P. tetrazona*.

Progeny Testing

In male heterogametic species, androgenesis results in the production of supermales (Y^2Y^2), which are academically and economically important animals. They are useful to understand the sex-determining mechanism in the tested species and to sire all-male progenies. Monitoring sex-ratios of progenies sired by supermales is a method to confirm the paternal integrity of the androgenotes. In male heterogametic species, the supermales are expected to sire all-male progenies (Devlin et al. 1991). Very few authors have extended their investigation to rearing the androgenotes to sexual maturity and assessing the sex-ratio of their progenies. Available publications clearly show the unexpected occurrence of 3-24% of F_1 female progenies sired by supermale. Kirankumar and Pandian (2004c) have induced successive generations of androgenotes (Fig. 4.6) and recorded the unexpected occurrence of females upto F_3 progenies, more or less in the same ratio, as it was in F_1 progenies. This investigation clearly indicates the need to identify the genotype of such unexpected female progenies.

Information on the sex-linked DNA markers in fishes is fragmentary and diverse. In a few fish species, different types of markers have been characterized; for instance, the Y-chromosome specific probe in the chinook salmon, *O. tshawytscha* (May et al., 1993) the sex chromosome specific repetitive sequences in the Poeciliids (Nanda et al., 1991), sex-specific quantitative DNA markers in *O. tshawytscha* (Clifton et al., 1997) and the male specific growth hormone pseudogene (GH-φ) in the masu salmon, *O. masou masou* (Zhang et al. 2001). These previous studies are related to members of Salmonidae and Poeciliidae. For the first time, Kirankumar et al., (2003) have identified,

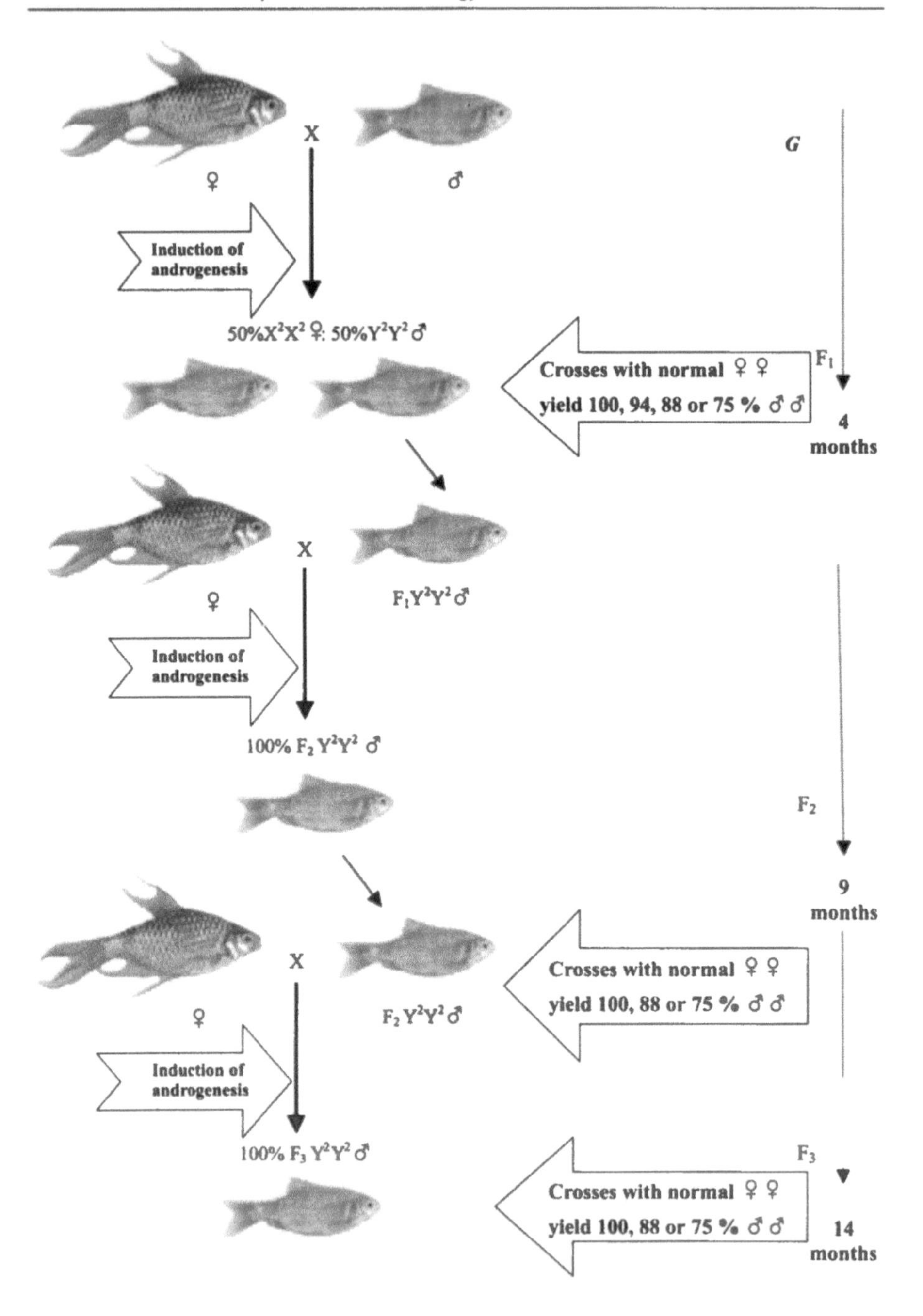

Fig 4.6: Protocol for production and progeny testing in androgenotic rosy barb

isolated and characterized a Y-chromosome specific molecular marker for the cyprinid *P. conchonius*. Using the SRY-specific primers, they have made PCR analysis of the genomic DNA of male golden rosy barb, which has yielded 3 amplicons of 588 bp, 333 bp and 200 bp length (Fig. 4.7). They have found that only 200 bp product is amplified in the female genome. Hence, the first 2 products may serve as molecular markers to rapidly identify a cyprinid fish possessing Y-chromosome. The consistent presence of the upper 333 and 588 bp fragments in normal (X^1Y^2), hormonally induced (Y^1Y^2) and androgenetic (Y^2Y^2) males and the absence of relationship between the 200 bp fragment and the X-chromosome clearly indicates that the male specific markers are specific to Y-chromosome only.

Incidentally, Kirankumar and Pandian (2003, 2004a,b) have alone reported that the reproductive performance of the androgenetic male is superior and the female inferior to that of the normal male and female, respectively (Table 4.7). If homozygosity is a cause for all the negative features recorded for the inferior reproductive performance of androgenetic female (X^2X^2), then it is difficult to comprehend the superior reproductive performance of androgenetic males (Y^2Y^2), which also possess an equal level of homozygosity.

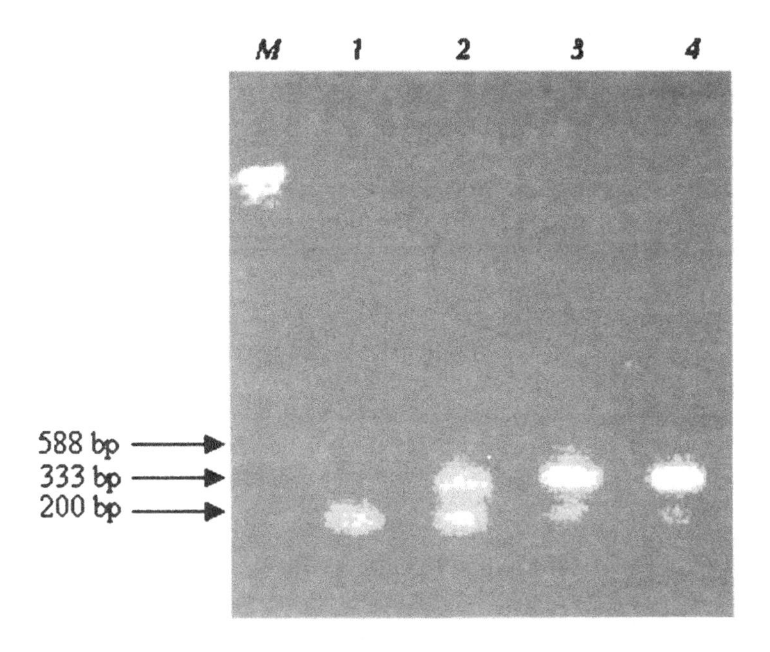

Fig. 4.7: PCR products amplified by *SRY* primers in the genomic DNA of the golden rosy barb, *Puntius conchonius* carrying different sex genotypes. Lane M – λ *Hind* III marker, Lane 1 – Normal female (X^1X^2), Lane 2 – Normal Male (X^1Y^2), Lane 3 – Hormonally induced supermale (Y^1Y^2), Lane 4 – Androgenetic supermale (Y^2Y^2) (from Kirankumar et al., 2003; Kirankumar and Pandian, 2004a)

Table 4.7: Reproductive performance of androgenetic males and females of the tiger barb, *Puntius tetrazona* and the rosy barb, *Puntius conchonius* (from Kirankumar and Pandian, 2003, 2004a)

Parametre	*Genotype of ♂ tiger barb*		*Genotype of ♂ rosy barb*	
	X^1Y^2	Y^2Y^2	X^1Y^2	Y^2Y^2
Sexual maturity (day)	110	120	90	85
GSI	0.48	0.52	0.38	0.55
Sperm count (no./ml)	8.1×10^5	8.8×10^5	7.2×10^5	7.5×10^6
Motility duration(s)	—	—	95	87
Fertilizability (%)	97	95	—	—
	Genotype of ♀ tiger barb		*Genotype of ♀ rosy barb*	
	X^1X^2	X^2X^2	X^1X^2	X^2X^2
Sexual maturity (day)	115	140	90	130
Inter-spawning period (day)	18	30	15	26
GSI	0.52	0.38	0.42	0.28

SURVIVAL OF ANDROGENETIC CLONES

In general, survival of androgenetic fishes is very low and the causes for which may be many but two are important: (1a) injury and stress involved in genome inactivation of egg by irradiation; (1b) stress imposed by thermal/pressure shock for diploidization and (2) homozygosity. Y and X-rays have high penetrance; hence they may cause damage not only to the genome but also to other important components of the eggs. Expectedly, the survival values reported for androgenotes generated using Y or X-rays for elimination of egg genome ranges from 0.7% to 12%. Conversely, the UV irradiation, known for its relatively lower penetrance, seem to cause less damage to other important components such as mRNA, mtDNA and proteins. Expectedly, the survival of these androgenotes—generated using eggs—in which genome was inactivated by UV irradiation, is higher and ranges around 15% (Table 4.2). Clearly, the UV irradiation protocol, first developed by Bongers et al., (1994) has not only removed the bottleneck of using expensive and skilled technique of Y and X-ray radiation but also improved by threefold the survival of the androgenotes.

Two different approaches have been made to eliminate the stress involved in diploidization; the first one is to use a relatively more homozygous diploid sperm while the second one involves the dispermic activation of eggs. In *O. mykiss*, Thorgaard et al., (1990) produced androgenotes using haploid and diploid sperm. Expectedly, viability of the androgenotes resulting from diploid sperm of a tetraploid male was significantly higher (43%) than those produced using haploid sperm (0.8%). Using diploid sperm, Arai et al. (1995) have also improved the survival of androgenotes of *M. anguillicaudatus* by almost 10 times. Clearly, the elimination of diploidization step in the protocol significantly increases the yield of androgenotes. Conversely, the survival of androgenotes arising from dispermic activation has not significantly improved; for instance, survival of this kind of androgenotes is reported as low as 0.1% in *O. mykiss* (Araki et al., 1995) and 3% for *P. conchonius* (Kirankumar and Pandian, 2004b). It is not clear why the androgenotes generated by dispermic activation result in such low survival, despite the fact that the entry of two different sperm should have considerably increased the heterozgosity.

At hatching, the survival is relatively higher and ranges from 0.7% in *O. mykiss* to 15% in *C. carpio* (Table 4.2). As they attain feeding stage, it decreases to 0.3% in *O. mykiss* (Thorgaard et al., 1990) and to 10% in *C. carpio* (Bongers et al., 1994). Thus, the survival of androgenotes is known to decrease with advancing age. For instance, it decreases from 15% at hatching to 7% at sexual maturity for *P. tetrazona* (Kirankumar and Pandian, 2003) and *P. conchonius* (Kirankumar and Pandian, 2004c). Apparently, about 84% of the mortality of the androgenotes occurs during embryonic development (Kirankumar and Pandian, 2004c). Kirankumar and Pandian (2004c) are perhaps the first to describe the stage-specific embryonic mortality and to pinpoint that the stages between activation and the 18th somite stage, and before and after hatching are critical stages. Interestingly, they also showed that: (i) despite activation by sperm belonging to another species, the development proceeds almost precisely in the same time scale as that of the normal egg fertilized by sperm belonging to same species (Fig. 4.8); and (ii) despite suffering severe mortality, a certain percentage of eggs, activated by sperm having the genome of related strain, regulates the normal development until hatching in the haploids but for an entire life time in the diploids.

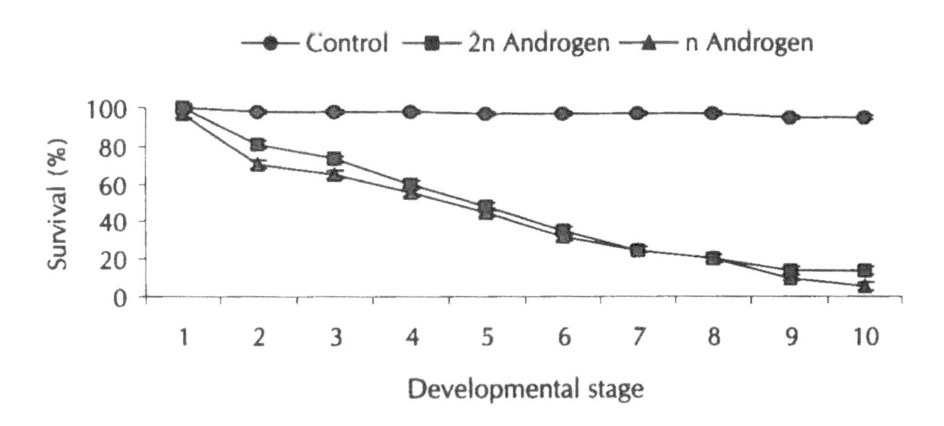

Fig. 4.8: Survival of androgenetic clone of golden rosy barb, *Puntius conchonius* as a function of the developmental stage. The Arabic numbers indicate selected embryonic stages (from Kirankumar and Pandian, 2004a)

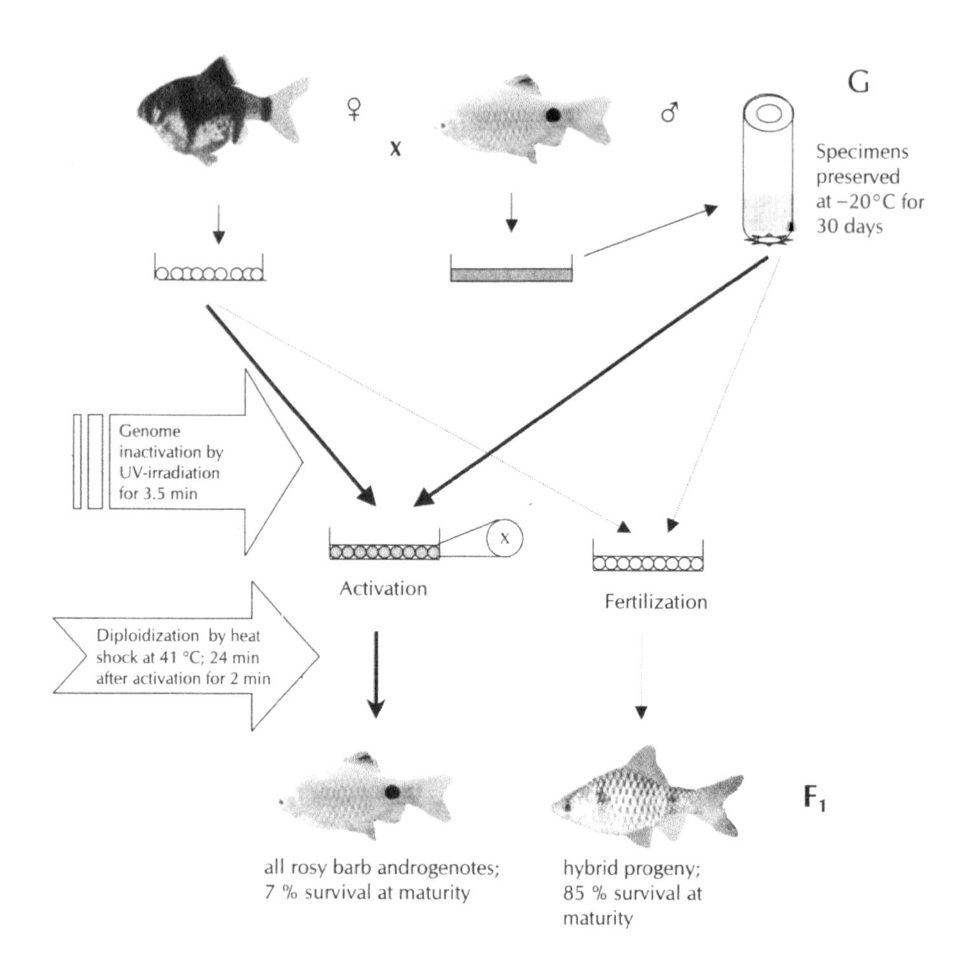

Fig. 4.9: Induction of interspecific androgenetic cloning of *Puntius conchonius* using its preserved sperm for activation of genome-inactivated (surrogate) eggs of *P. tetrazona* (from Kirankumar and Pandian, 2004c)

In the context of heterospecific insemination and intergeneric androgenetic cloning, it is interesting to note the attempt made by Kirankumar and Pandian (2004a) to understand the role played by haploid, diploid or hybrid genome in regulation of time sequence of development and embryonic mortality in *P. conchonius* (egg size: 882 µm) and *P. tetrazona* (egg size: 1230 µm), characterized by differences in egg size, and pre- and post-embryonic durations (Table 4.8). They made the following conclusions: (i) the haploid genome regulates the developmental sequence as effectively as that of the diploid genome; (ii) the haploid or diploid rosy barb genome, drawn from one or two fresh sperm or one cadaveric sperm, regulates the developmental sequence, even in alien surrogate egg of tiger barb, as good as that of the diploid genome of the tiger barb; (iii) the time sequence characteristic of the rosy barb is maintained, even when about 25% excess yolk was available in eggs of the tiger barb; and (iv) hybrid eggs and haploids suffer heavy mortality.

INTERSPECIFIC CLONING

An alternate approach to increase the heterozygosity is to generate interspecific androgenotes. Although interspecific androgenesis has been attempted in many species e.g. Salmonidae (May and Grewe, 1993); Cyprinidae (Grunina et al., 1995; Bercknyski et al., 1998) were the first to produce viable androgenetic goldfish, C. *auratus* using its fresh or frozen sperm to activate the genome-inactivated eggs of the common carp, C. *carpio* (Table 4.1). Our specific understanding of nucleo-cytoplasmic relation in fish is fragmentary. Chinese scientists like Prof. T.G. Tung presumed the enucleation of eggs of crucian carp, C. *auratus*, by pricking a glass microneedle at the point immediately underneath the very small polar body. Subsequently, the nucleus obtained from the blastula cells of common carp was introduced into the presumed enucleated eggs of crucian carp. This kind of transplantation helped generate hybrids between these cyprinids. Briefly, such studies indicate that both nucleus and cytoplasm influence the expression of genetic information in the hybrid. Barring these Chinese publications, there is no readily available

Table 4.8: Duration required for completion of pre- and post-embryonic development and survival of the tiger and rosy barb, their hybrids and rosy barb androgenotes. (from Kirankumar and Pandian, 2003, 2004a, b, c)

Cross	*Duration from fertilization to hatching (hr)*	*Duration from hatching to feeding (hr)*	*Survival (%) at hatching*	*maturity*
Tiger barb control (TB)	26	42	98	94
Rosy barb control (RB)	24	36	98	95
RB ♀ x TB ♂ - hybrid	24	36	72	0
TB ♀ x RB ♂ - hybrid	24	40	80	75
2n interspecific androgenote (fresh RB sperm)	24	42	14	7
2n interspecific androgenote (cadaveric RB sperm)	24	42	7	3
2n interspecific androgenote (fresh RB n + n sperm)	24	42	3	2
n interspecific androgenote (fresh sperm source)	24	0	9	0
n interspecific androgenote (cadaveric sperm source)	24	0	3	0

publication in English literature on the nucleo-cytoplasmic relationship in eggs and hybrid eggs of fishes (see Pandian, 2002). Until adequate information is made available on hybrid eggs, protocols for generation of interspecific androgenotes will have to be made on trial and error basis. To make an interspecific androgenetic cloning a success, the following issues are taken into consideration, when selection for the 2 species is made: (i) compatability between the yolk volume and incubation period; (ii) compatibility of the head of the donor sperm and the micropyle of the host egg; (iii) availability of the phenotypic and protein markers as well as species-specific primers for PCR analysis; and (iv) affording protection to 'the maternal produts' of the eggs namely enzyme, mRNA and mtDNA while the eggs are being irradiated

Table 4.9: Source, techniques and genome of the clones generated in mammals and fishes

Particular	*Mammals*	*Fishes*
Source	Enucleated ovum of recipient and nucleus of a highly differentiated cell of the donor	Due to non-visiblity of nucleus, genome-inactivated (by UV irradiation), egg as recipient and sperm as the donor
Requirement	Viviparity requires a surrogate mother	With oviparity, the genome-inactivated egg replaces the requirement of surrogation
Techniques	Highly skilled technique of enucleation of recipient egg and donor cell	Not so skilled technique of UV irradiation to inactivate egg's genome
	Cell fusion technique to transfer donor's genome	Natural sperm-mediated transfer of donor's genome
	Surgical introduction of 'fused' egg in surrogate mother	Interspecific androgenetic cloning requires dispermy or diploid sperm, or monospermy or haploid sperm with restoration of diploidy by thermal/pressure shock
Clone's genome	Originating from a highly differentiated cell; clone develops in egg's cytoplasm, i.e. "old wine in a new bottle"	Originating from donor's sperm, the clone develops in the egg's cytoplasm, i.e. "new wine in a new bottle"
	Nuclei of different cells of the same tissue may generate clones like 'xerox copies' of the donor, i.e. clones may be identical and may not differ genetically from each other	Nuclei of different sperm may generate clones like 'photos' of the donor, i.e. clones may be identical but may genetically slightly differ from each other
Clone's sex	Since the clones originate from the same nuclei, say, udder cells of a donor, they are all expected to be of the same sex, i.e. females only	Since the clones originate from nuclei of either X or Y carrying sperm, expected sex ratio of clones 50% females and 50% males
Advantage/ limitation	Requires live donor and recipient	Requires live recipient but the donor can be from the sperm of either live or post-mortem preserved donor

so as to inactivate their genome. These technical difficulties may prove to be a tough task but are not totally insurmountable.

Not surprisingly, interspecific androgenetic clones have been generated only in a couple of species, namely *C. auratus* using *C. carpio* eggs (Bercsenyi et al., 1998) and *P. conchonius* using *P. tetrazona* eggs (Kirankumar and Pandian, 2004a; Fig. 4.9). Kirankumar and Pandian (2004a) have successfully reared the interspecific androgenetic clones to sexual maturity and found that the reproductive performance of these androgenetic males was superior and that of the females as inferior to that of the respective controls (Table 4.7). More recently, intergeneric androgenetic clones of *Hemigrammus caudovittatus*, using its preserved sperm and genome-inactivated eggs of *Gymnocorymbus ternetzi*, have successfully been generated (David and Pandian, 2004).

Among vertebrate groups, fish are uniquely amenable for induction of interspecific androgenotes. Indeed, techniques and protocols for induction of interspecific and intergeneric androgenotes will lay the foundation for : (i) restoration of endangered fish using preserved/cadaveric sperm and surrogate egg; (ii) to produce seedlings almost throughout the year in annual spawners like the carps using surrogate eggs of undesired fish like tilapia, which breeds throughout the year; and (iii) generation of seedlings in migratory species like eel and hilsa using their cryopreserved sperm and suitable surrogate eggs.

Despite its comparability, androgenetic cloning is not identical to the mammalian cloning achieved in recent years. Table 4.9 presents a comparative account on the source, techniques and genomes used for generation of clones of mammals and androgenetic clones of fish. From the points of conservation and aquaculture, androgenetic cloning certainly has an edge over the technic of mammalian cloning (e.g. Wilmut et al. 1997).

ACKNOWLEDGEMENTS

We gratefully acknowledge Prof. O. Kinne for his constant encouragement and the Council of Scientific and Industrial Research and Indian Council of Agricultural Research, New Delhi, for financial support.

REFERENCES

Afzelius, B.A., 1978. Fine structure of the garfish spermatozoan. J. Ultra. Res., 64: 309-314.

Arai, K., Onozato, H. and Yamazaki, F., 1979. Artificial androgenesis induced with gamma irradiation in masu salmon, *Onchorynchus masou*. Bull. Facult. Fish., Hokkaido Univ. 30: 181-186.

Arai, K., Ikeno, M. and Suzuki, R., 1995. Production of androgenetic diploid loach *Misgurnus anguillicaudatus* using spermatozoa of natural tetraploids. Aquaculture, 137: 131-138.

Araki, K., Shinma, H., Nagoya, H., Nakayama, I. and Onozato, H., 1995. Androgenetic diploids of rainbow trout (*Onchorynchus mykiss*) produced by fused sperm. Can. J. Fish. Aquat. Sci., 52: 892-896.

Ayyappan, S., Ponniah, A.G., Reddy, P.V.G.K., Jana, R.K., Mahapatra, K.D. and Basavaraju, Y., (eds. Gupta, M.V. and Acosta, B.O.) 2001. Aquaculture genetics research in India: An

Palmer, P.J., Blackshaw, A.W. and Garrett, R.N. 1993. Succesful fertlity experiments with cryopreserved spermatozoa of barramundi *Lates calcarifer* (Bloch) using dimethyl sulfoxide and glycerol as cryoprotectants. Reprod. Fertil. Dev., 5: 285-293.

Pandian, T.J. 2002. Cloning the fish. Curr. Sci., 83: 1063-1064.

Pandian, T.J. and Koteeswaran, R. 1998. Ploidy induction and sex control in fish. Hydrobiologia, 384: 167-243.

Parsons, J.E. and Thorgaard, G.H. 1985. Production of androgenetic diploid rainbow trout. J. Hered., 76: 177-181.

Ponniah, A.G., Kapoor, R.N. and Lal, K.K. 2001. Preliminary investigation on androgenesis in *Cyprinus carpio*. Natl. Acad. Science Letters, 23.

Romashov, D.D. and Belyaeva, V.N. 1964. Cytology of radiation gynogenesis and androgenesis in the loach. Dokl. Akad, Nauk. SSSR., 157: 964-984.

Rosenthal, H. and Odense, P. 1989. Fertlization in Herring eggs: A scanning electron microscopic study. In: *"Fish Health and Protection Strategies"*, (eds) K. Lillielund and H. Rosenthal Federal Ministry for Research and Technology, Hamburg, pp. 199-205.

Scheerer, P.D., Thorgaard, G.H., Allendorf, F.W. and Knudsen, K.L. 1986. Androgenetic rainbow trout produced from inbred and outbred sperm sources show similar survival. Aquaculture, 57: 289-298.

Scheerer, P.D., Thorgaard, G.H. and Allendorf, F.W. 1991. Genetic analysis of androgenetic rainbow trout. J. Exp. Zool., 260: 382-390.

Scott, A.P. and Baynes, S. M. 1980. A review of the biology, handling and storage of the salmonid spermatozoa. J. Fish Biol., 17: 707-739.

Stroband, H.W.J., Te Kronnie, G. and van Gestel, W. 1992. Differential susceptibility of early steps in carp (*Cyprinus carpio*) development to α-amantin. Roux's Arch. Develop. Biol, 202: 61-65.

Stoss, J., Fish gamete preservation and spermatozoan physiology. In: Fish Physiology (eds) Hoar, W.S., Randall, D.J. and Donaldson, E.M., Academic Press, New York, 9B, 305-350.

Stanley, J.G., Biggers, C.J. and Schultz, D.E. 1976. Isozymes in androgenetic and gynogenetic white amur, gynogenetic carp and carp-amur hybrids. J. Hered., 67: 129-134.

Thorgaard, G.H., Scheerer, P.D., Hershberger, W.K. and Myers, J.M. 1990. Androgenetic rainbow trout produced using sperm from tetraploid males show improved survival. Aquaculture, 85: 215-221.

Varadaraj, K. 1990. Production of diploid *Oreochromis mossambicus* gynogens using heterologous sperm of *Cyprinus carpio*. Ind. J. Exp. Biol., 28: 701-705.

Ueda, T., Kobayashi, M. and Sato, R. 1986. Triploid rainbow trouts induced by polyethylene glycol. Proc. Jap. Acad., 62B, 161-164.

Wilmut, I., Schnieke, A.E., McWhir, J., Kind, A.J. and Campbell, K.H. 1997. Viable offspring derived from fetal and adult mammalian cells. Nature, 385: 810-813.

Wu, C., Chen, Y.R. and Liu, X. 1993. An artificial multiple triploid carp and its biological characteristics. Aquaculture, 111: 255-262.

Zhang, Q., Nakayama, I., Fujiwara, A., Kobayashi, T., Oohara, I., Masaoka, T., Kitamura, S. and Devlin, R.H. 2001. Identification by male-specific growth hormone pseudogene (GH-Ψ) in *Onchorynchus masou* complex and a related hybrid. Genetica, 111: 111-118.

Chapter 5

Methods of Sex Control in Fishes and an Overview of Novel Hypotheses concerning the Mechanisms of Sex Differentiation*

C.A. Strüssmann,[1] M. Karube[1], L. A. Miranda[2], R. Patiño[3], G. M. Somoza[2], D. Uchida[4,5], and M. Yamashita[5]

[1] Department of Marine Biosciences, Faculty of Marine Science, Tokyo University of Marine Science and Technology. Konan 4-5-7, Minato, Tokyo 108-8477, Japan, Email: carlos@s.kaiyodai.ac.jp

[2] Instituto de Investigaciones Biotecnológicas / Instituto Tecnológico de Chascomús (CONICET-UNSAM). Camino de Circunvalación Laguna, Km 6. cc 164, (7130) Chascomús, Provincia de Buenos Aires, Argentina

[3] Texas Cooperative Fish & Wildlife Research Unit / Texas Tech University, Lubbock, TX 79409-2120, U.S.A.

[4] National Research Institute of Fisheries Science, Fukuura, Kanazawa, Yokohama 236-8648, Japan

[5] Graduate School of Integrated Science, Yokohama City University, Seto, Kanazawa, Yokohama 236-0004, Japan

ABSTRACT

This chapter discusses the methods of sex control and current knowledge of the mechanisms of gonadal sex differentiation in fishes. Although our understanding of sex differentiation in fishes has improved greatly in recent years, it is still inadequate either to allow generalizations about its mechanisms or to develop practical, reliable and safe methods of sex control. The scarcity of information arises from the fact that studies on sex differentiation are generally conducted on a few traditional species and focus on a limited set of hypotheses. Therefore, the current knowledge does not account for the variation in the forms of sexual expression in fishes and for the various degrees of plasticity of phenotypic sex that are found among fishes. In order to broaden the discussion about the

*All authors are equally credited for this publication. Correspondence should be sent to CAS

mechanisms of sex differentiation, we present here an overview of our studies on the sex differentiation of three species, the pejerrey (*Odontesthes bonariensis*), the channel catfish (*Ictalurus punctatus*), and the zebrafish (*Danio rerio*), and introduce some novel views about this process. The hypotheses being explored are that: (a) temperature-dependent sex determination/differentiation is under the control of the central nervous system; (b) differential sex-linked distribution of cells expressing the estrogen receptor plays a role in early gonadal sex differentiation; and (c) apoptosis is involved in the death of somatic or germ cells during sex differentiation.

Keywords: Fishes, sex differentiation, sex control, environmental sex determination, apoptosis, estrogen receptor, aromatase, gonadotropin

SEX CONTROL AND GONADAL SEX DIFFERENTIATION

The ability to control sex phenotype and reproductive activity in fishes is of benefit to aquaculture (Hunter and Donaldson, 1983; Thorgaard, 1983; Patiño, 1997; Pandian et al., 1999). For example, the suppression of reproduction is fundamental to increase productivity, as unrestrained reproduction during grow-out often leads to wastage of energy in the form of gamete production and reproductive behaviour that could otherwise be channelled to somatic growth. Second, in a number of species, one sex grows faster or has better survival than the other, or has certain other characteristics that appeal to particular markets. The eggs of species such as sturgeon, mullet and salmon, or gravid females such as ayu (smelt) fish are considered a delicacy in some countries. Likewise, males of guppy and many other species are highly valued in the market of ornamental aquarium fishes. The ability to control reproductive activity can be of benefit not only to commercial fish culture operations, but also to fish management practices. For example, fishes released into the environment to support recreational fisheries or to control weeds or pests could impact the aquatic ecosystem unless their reproductive activity can be controlled. In some cases, exotic species have caused profound and even irreversible ecological changes such as the extinction of autochthonous species and the loss of biodiversity (De Silva, 1989; Moyle, 1991; Carlton, 1992; Leach, 1995; Lodge et al., 1998). Finally, sex control is obviously required in the case of genetically engineered organisms to avoid the introduction of modified genes into the native gene pool (Bartley and Hallerman, 1995; Hallerman and Kapuscinski, 1995; Pandian et al., 1999).

The methods most commonly employed to manipulate the sex of fish are based on exogenous hormone treatment, chromosome manipulation or a combination of both (Hunter and Donaldson, 1983; Thorgaard, 1983; Pandian et al., 1999). It is beyond the scope of this discussion to cover the principles, advantages and limitations of each of these approaches. For this purpose, the reader is referred to the recent elegant review of Pandian et al. (1999). In spite of its obvious potential benefits, however, practical utilization of sex control in fishes is still very limited. Reasons for the relative lack of application of this technology include the low viability (survival or growth) of the treated population (specially after chromosome manipulation), the production of populations that are not completely monosexual (male or female), and concerns over health and environmental hazards (especially with techniques involving hormone manipulation) (Pandian et al. 1999). Concerning the production of monosex populations, the single major reason underlying the lack of reliability of current technology is the plasticity (bipotentiality) of gonadal sex, which in many species is evident even in adult individuals (Redding and Patiño, 1993; Patiño, 1997; Nakamura et al., 1998; Pandian and Koteeswaran, 1999; Pandian et al., 1999). However, in many species, the sex is determined early in life and remains stable thereafter (Patiño, 1997; Strüssmann and Patiño, 1999). Thus, in these latter species, the major requirement to develop reliable and safe sex control protocols is improved knowledge of the mechanisms of sex differentiation (including genetic determinants) and the effects of exogenous (chemical or environmental) factors on this process.

Our understanding of the molecular and cellular mechanisms of sex differentiation in fishes has increased exponentially over the last several decades. A number of critical steps involved in this process and some of the specific regulatory genes have been identified (see reviews by Yamamoto, 1969; Hunter and Donaldson, 1983; Nakamura et al., 1998; Baroiller et al., 1999; Strüssmann and Nakumara, 2002). The majority of the literature seems to support the concept promoted by Yamamoto (1969) that natural steroid hormones are key regulatory factors in the cascade of events associated with gonadal sex differentiation (Adkins-Regan, 1987; Redding and Patiño, 1993; Patiño, 1997; Nakamura et al., 1998; but also see conflicting views in Kawahara and Yamashita, 2000; Strüssmann and Nakumara, 2002). Furthermore, steroidogenic enzymes and steroid receptors are elements that can be influenced by environmental temperature (Crews et al., 1994; Baroiller et al., 1999), and this could explain the growing number of reports of temperature-dependent sex determination among fishes (Strüssmann and Patiño, 1995, 1999; Pandian and Koteeswaran, 1999). Notwithstanding these recent advances in knowledge, our understanding of the mechanisms of sex differentiation in fishes is far from complete. Past experience has demonstrated the risk of making generalizations based on observations made with only a handful of species. Such generalizations may be possible only when information is available for species representing the full spectrum of the forms of sexuality found in fishes. Fishes include species that reproduce gonochoristically, hermaproditically (sequential or simultaneous), and parthenogenetically (unisexual) (Yamamoto, 1969). Moreover, sex differentiation among gonochorists can proceed directly from the undifferentiated gonads into a male or female pathway ("differentiated gonochorists") or through an initial female phase followed by sex reversal into males in half of the individuals ("undifferentiated gonochorists") (Yamamoto, 1969; Nakamura et al., 1998).

Finally, the functional development of sex in a species may follow strictly a predetermined (supposedly genetic) pathway or be labile to the action of physical, chemical, or social cues (Korpelainen, 1990; Francis, 1992; Strüssmann and Patiño, 1995, 1999; Pandian and Koteeswaran, 1999).

This chapter is an overview of our studies on the sex differentiation of three fish species, the pejerrey (*Odontesthes bonariensis*, order Atheriniformes), the channel catfish (*Ictalurus punctatus*, order Siluriformes), and the zebrafish (*Danio rerio*, order Cypriniformes) and—for the purpose of stimulating discussion—to introduce novel hypotheses regarding this process. The hypotheses being explored are that: (a) temperature-dependent sex determination/differentiation is under the control of the central nervous system; (b) differential sex-linked distribution of cells expressing the estrogen receptor (ER) plays a role in early gonadal sex differentiation; and (c) apoptosis is involved in the death of somatic or germ cells during sex differentiation. The species under study have relatively strong genotypic (zebrafish and channel catfish) or environmental (pejerrey) mechanisms of sex determination, and represent also the two forms of sex differentiation found among gonochoristic teleosts: the undifferentiated (zebrafish) and differentiated (channel catfish and pejerrey) types of gonads.

TEMPERATURE AND SEX DIFFERENTIATION

In several fish species, the phenotypic sex can be determined by the water temperature during a critical period early in life. This phenomenon is known as thermolabile sex determination (TSD) (Korpelainen, 1990; Francis, 1992; Strüssmann and Patiño, 1995, 1999; Pandian and Koteeswaran, 1999). The mechanisms involved in thermolabile sex determination are still not clearly understood. In TSD of reptiles, gonadal steroids appear to be the physiological transducers of the incubation temperature (Crews, 1996; Pieau, 1996) and such a link is now being investigated in teleost fishes (Baroiller et al., 1999; D'Cotta et al., 1999; Kitano et al., 1999; Strüssmann and Nakamura, 2002). At present, it is assumed that in fishes the central nervous system does not play a preponderant role in the TSD process or in the process of sex differentiation in general (Baroiller et al., 1999). However, treatment with pituitary extracts accelerated gonadal and germ cell development in larval carp, *Cyprinus carpio* (Van Winkoop et al., 1994). Also, Fitzpatrick et al., (1993) noted a stimulation of the synthesis of certain steroids in the anterior kidney of sexually undifferentiated rainbow trout after treatment with salmon gonadotropins (GTHs). Finally, the brain-pituitary axis is the likely start for any mechanism that depends on the perception of external cues such as environment-induced primary sex determination or sex reversal (Francis, 1992).

The pejerrey (*O. bonariensis*) is a differentiated gonochorist, at least at intermediate temperatures (Strüssmann et al., 1996a), and presents marked TSD (Strüssmann et al. 1996b, 1997). In this species, the proportion of females changes gradually from 100% at 15-19°C to 0% at 29°C, and the critical time of temperature-dependent sex determination was estimated to be between 3-5 weeks, 2-4 weeks, and 1-4 weeks after hatching at 17°C, 19°C and 27°C, respectively (Strüssmann et al., 1997). Moreover, experiments to date involving over 15 different broods have indicated a very consistent response to the extreme temperatures (17°C and 29°C) (Strüssmann et al. 1996b, 1997). These facts make pejerrey an ideal model to study the effects of temperature on the processes of sex determination and differentiation. We are currently conducting a comprehensive study to examine the local (gonadal) phenomena that occur during gonadal sex differentiation and the possibility of their regulation by the hypothalamo-pituitary axis. In the ongoing studies, the gonads and pituitaries of fish reared from hatching to 11 weeks of age at feminizing (17°C), intermediate (24°C), and masculinizing temperatures (29°C) were examined on a weekly basis by histological, immunocytochemical, stereometrical or molecular biological methods. The following is a brief synopsis of the results of these analyses; detailed accounts will be published elsewhere.

Histological differentiation of the gonads was analysed based on the criteria established by Strüssmann et al. (1996a). The proportion of females in groups reared at 17°C, 24°C, and 29°C was 100%, 78%, and 0%, respectively. Sex differentiation in males (24°C and 29°C) and females (17°C and 24°C) followed the same basic histological pattern regardless of water temperature. These results confirmed the fact that sex differentiation is of the "differentiated" mode (Strüssmann et al. 1996a). In other words, the gonads develop directly into either sex even at the temperatures that induce 100% males or females, without the degeneration of temporary female elements that typifies the process of testicular formation in species that follow the "undifferentiated" mode. Nevertheless, in recent studies, we have observed a higher percentage of degenerating germ cells with increasing temperatures during the period of gonadal sex differentiation. This phenomenon has been reported previously only for fish at the juvenile stage (see Strüssmann et al., 1998; Strüssmann and Patiño, 1999; Ito et al., 2003) and its relevance for the process of sex differentiation in this species is currently being studied.

Ovarian differentiation was first recognized at week 7 (individual range of 6-8 weeks; minimum and maximum body lengths at sex differentiation of 19.8 and 27.4 mm, respectively) at 17°C and week 4 (individual range of 4-5 weeks; minimum and maximum body lengths of 16.5 and 21.4 mm, respectively) at 24°C. Testicular differentiation began at week 7 (only one male out of 10 individuals; body length of 29.4 mm) at 24°C and at week 4 (maximum and minimum body lengths of 17.7 and 19.7 mm, respectively) at 29°C. Blood vessels appeared at 6, 3-4 (same time for putative males and females), and 3 weeks at 17°C, 24°C, and 29°C, respectively. Thus, blood vessels were formed in all the individuals shortly before or concomitant with the first signs of commitment to either sex at all temperatures. Similar findings have been reported in tilapia, *Oreochromis niloticus* (Nakamura et al., 1998). Interestingly, the relative size of the blood vessels in relation to the sectional area of the ovary was larger than that of the presumptive testes from the first day of appearance. Thereafter, the relative area of the blood vessels remained constant in the testes but continued to grow larger in ovaries, widening even further the sexual dimorphism of the gonads. The observations that blood vessels were formed before or at the time of commitment of the gonads to a male or female pathway of development, and the existence of a morphological sex dimorphism at this early stage, are consistent with the concept of extragonadal regulation of sex differentiation in this species (see below).

Molecular biological analysis of sex differentiation in pejerrey aims to verify whether differential expression of aromatase, an enzyme that converts androgens into estrogens, and ER genes (α and β) can explain the induction of testes and ovaries at high and low temperatures, respectively, as have been suggested for many fish and reptile species. Thus, we have cloned the complete sequence of pejerrey ovarian aromatase, including the open reading frame and have obtained partial sequences for ovarian ERα and ERβ, and also estimated the level of gene expression by semiquantitative RT-PCR. Preliminary results of these analyses indicate that aromatase, ERα and ERβ expression begins before or around the time of appearance of the first signs of histological sex differentiation of the gonads in pejerrey. The expression of aromatase apparently follows a sexually dimorphic pattern regardless of temperature: it gradually increases in presumptive females, whereas in males the expression was low but constant throughout. Considering only the values at the extreme temperatures, these results agree with other publications in that aromatase gene expression is weaker at male- than at female-forming temperatures (Baroiller et al., 1999; Kitano et al., 1999). Unfortunately, neither of these studies has been able to clarify whether differential aromatase expression precedes or merely follows the commitment of the gonads to either sex. In pejerrey, increased aromatase expression occurs before and after the onset of histological sex differentiation of females at 17°C and 24°C,

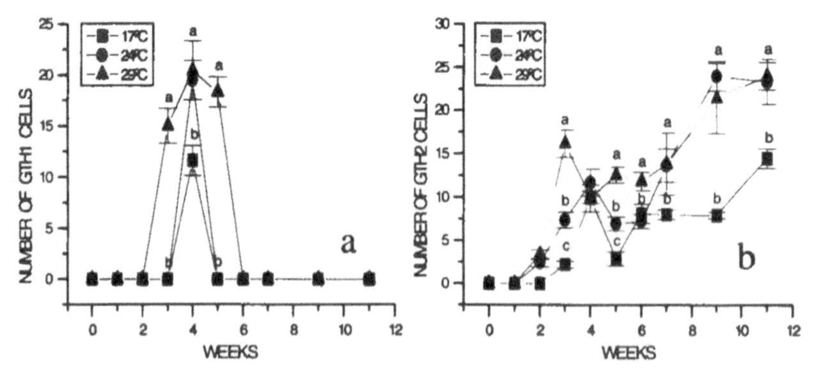

Fig. 5.1: Weekly changes in the number of immunoreactive GTH1 (a) and GTH2 (b) cells in the proximal *pars distalis* of *O. bonariensis* reared from hatching to 11 weeks at 17°C, 24°C, or 29°C. Each bar represents the mean of 4 animals ± SEM. Different letters indicate statistically significant differences between groups ($P < 0.05$). (Reprinted with permission from Miranda et al. (2001).

respectively. This observation, associated with the fact that expression levels at the temperature in which both sexes were formed (24°C) were bimodal rather than being intermediate between those at the lowest and highest temperatures, seems to support the concept that aromatase expression follows commitment to ovarian differentiation. Interestingly, the expression levels of ERα increased with development regardless of temperature, whereas the ERβ did not follow any particular pattern of expression during the same period. Ongoing studies should confirm these preliminary results and examine the regulatory mechanisms of these genes, possibly by the central nervous system (see below), as well as whether the differential gene expression observed in this study for aromatase actually translates into differential enzyme or ER binding activity. Also, the possibility of the existence of two aromatase genes in pejerrey—as has been recently suggested for other species—and their differential roles during sex differentiation will need to be clarified.

We used immunocytochemistry to investigate the ontogeny of GTH expression during the temperature-sensitive period of sex determination (Miranda et al., 2001). We focussed on GTH cells based on the results of Fitzpatrick et al. (1993) and because studies addressing the ontogeny and the roles of GTHs in early gonadal development (Van den Hurk, 1982; Mal et al., 1989; Feist and Schreck, 1996) revealed a marked species-specific variation in the timing of appearance of the two forms of GTH, 1 and 2 (structurally homologous to tetrapod FSH and LH, respectively). Our present analysis revealed two distinct types of gonadotropic cells in the pejerrey pituitary showing either GTH1-ß or GTH2-ß immunoreactivity, as has been previously demonstrated in other teleosts (Nozaki et al., 1990a,b; Toubeau et al., 1991; Lin et al., 1992; Kagawa et al., 1998; Yoshiura et al., 1999). The GTH1-immunoreactive cells were found mainly in the anterior part of the proximal *pars distalis* (PPD) close to the entrance of the neurohypophysis whereas the GTH2-expressing cells were observed in the posterior part and the external layer of the PPD and in the *pars intermedia*.

Both types of GTH cells were identified before sexual differentiation in pejerrey (Miranda et al., 2001). The GTH1 cells were first recognized at week 4 for larvae kept at 17°C and 24°C and at week 3 for larvae kept at 29°C. During the experimental period, GTH1 cells showed a clear peak in number at week 4 at all temperature regimes but disappeared completely after week 6 and until the end of the experiment (Fig. 5.1a). The number of GTH1 cells was statistically higher at 29°C when compared either to 24°C or 17°C. In the case of GTH2, the first cells were observed after 3 weeks in larvae at 17°C and 2 weeks at 24°C and 29°C. The number of GTH2 cells showed a transient peak at week 4 in larvae held at 17°C and 24°C and at week 3 for those held in 29°C, then continued to increase until the end of the experiment at 11 weeks (Fig. 5.1b). Interestingly, GTH2 cells appeared earlier and were always more numerous than GTH1 cells (Fig. 5.1). Thus, peaks in the number of immunoreactive cells were observed shortly before the first histological signs of gonadal sex differentiation at all temperatures and at a time when the gonads are probably most sensitive to temperature (Strüssmann et al., 1997). The magnitude and timing of the appearance of these peaks varied with temperature, suggesting that GTH1 and/or GTH2 may be important for either sex determination or differentiation in pejerrey. This assumption is consistent with the observation that GnRH neurons and fibres can be found in the pejerrey brain before sexual differentiation and concomitantly with the appearance of GTH cells (Miranda et al., 2002). One hypothesis that could be formulated based on these results is that gonadotropins are released from the pituitary and reach the gonads or interrenal tissue through the general circulation, which as shown above, is already functional at this stage, where they stimulate steroidogenesis (Fitzpatrick et al. 1993) and promote gonadal sex differentiation and growth. Ongoing research should help clarify the involvement of GTHs and other brain and pituitary hormones in the early differentiation of the gonads of pejerrey and other species.

SEX DIFFERENTIATION AND ER-IMMUNOREACTIVE GONADAL CELLS

Few studies have examined the possibility that changes in ER expression participate in the process of gonadal sex differentiation in fishes. The results of these studies seem to indicate that there are no considerable differences in ER gene (RNA) expression between genetic males and females (Guigen et al., 1999) or between individuals raised at female- or male-inducing temperatures (pejerrey, this study), although an increase in ERα expression may have occurred simultaneously with the onset of differentiation in pejerrey ovaries (see earlier discussion). However, ER genes may be expressed in several potentially relevant tissues besides the gonads, such as liver (e.g. Pakdel et al., 1989), brain (e.g. Salbert et al., 1991; Hawkins et al., 2000), and trunk kidney (Xia et al., 2000; Patiño et al., 2000). Also, assays relying on whole-tissue (Guigen et al., 1999) or whole-body (pejerrey, present study) extracts to determine gene expression may fail to detect relatively small, compartmentalized changes in expression. Thus, it is also necessary to identify the precise tissues and cells that express the genes of interest, and preferably the presence of the protein rather than RNA should be measured. In this study, we have examined the presence of cells expressing ERα-immunoreactive protein during sex differentiation in channel catfish (*Ictalurus punctatus*). Channel catfish is a "differentiated" gonochorist in which the gonadal sex is determined largely by the genotype. However, high temperatures during the critical time of sex differentiation can override the genetic control in channel catfish and induce the formation of a higher proportion of females in the population (Patiño et al. 1996).

For the present analysis, a rabbit antiserum was raised against an epitope of the D-domain of the channel catfish ERα (Xia et al. 1999). This antiserum (but not pre-immune serum) reacted positively with cells of the trunk kidney of catfish fry (Fig. 5.2). The structure of these cells resembled that of lymphopoietic cells in the kidney of older fish.

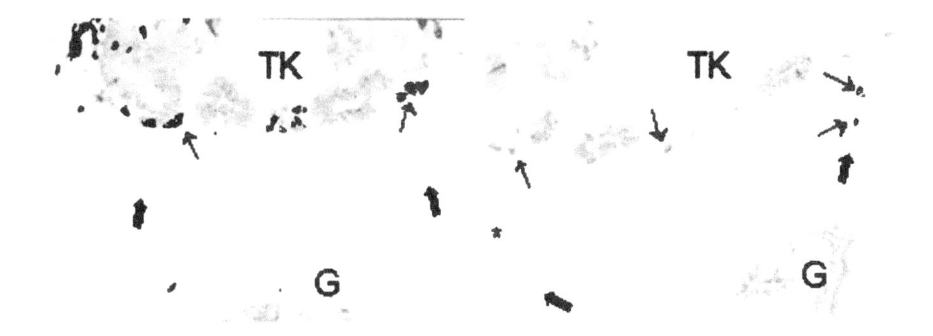

Fig. 5.2: Cross-sections of the trunk region of (a) genetic male and (b) sex-reversed female (genetic male) channel catfish treated with feminizing steroid (see Patiño et al., 1996). Both groups of fish were sampled 19 days after fertilization. The testes of the fish sampled were sexually indifferent but the ovaries showed early morphological signs of differentiation such as slightly larger sizes and onset of tissue outgrowths (asterisk). Thin arrows point to some of the ERα-immunoreactive cells (dark appearance) present in trunk kidney (TK) or gonads, and block arrows point to the gonads. G, gut.

The trunk kidney in fishes is found parallel and in close proximity to the indifferent gonads along the upper wall of the abdominal cavity, from just behind the air bladder to the posterior end of the abdomen. The close physical association between the kidney and the primordial gonads and the early development of the steroidogenic ability of the kidney indicate a functional association between the two tissues (e.g. Fitzpatrick et al. 1993). Interestingly, the ovaries of sex-reversed females (genetic males treated with feminizing steroid) also contained ERα-immunoreactive cells (Fig. 5.2b). These cells first appeared in the presumptive ovaries shortly before the onset of gonadal sex differentiation (which starts about 19 days post-fertilization; Patiño et al., 1996). The cytological appearance of the cells found in the ovary resembled that of the cells from the kidney (Figs 5.2a, b). However, ERα-immunoreactive cells were rarely seen in presumptive testes of genetic males of the same age (Fig. 5.2a). In all ERα-immunoreactive cells of the kidney and gonad, the antigen recognized by our antiserum seemed to be primarily localized in the cytosol. The expression of ERα RNA in lymphopoietic tissue of the trunk kidney of older catfish has been recently confirmed by in situ hybridization (Patiño et al. 2000).

The biological significance of the present observations is uncertain. Further studies on the cytosolic antigen recognized by our catfish anti-ERα antiserum are required before the significance of these observations can be fully understood. However, the differential distribution of ER α-immunoreactive cells during gonadal sex differentiation (higher number in the differentiating ovary than in the presumptive testis) suggests a role for these cells in the process of sex differentiation. Ongoing studies aim to quantify the apparent difference in the number of ERα-immunoreactive cells between male and female gonads, to determine the tissue source and nature of the gonadal ERα-immunoreactive cells, and to examine the presence of these cells in differentiating ovaries of other fish species.

APOPTOSIS AND SEX DIFFERENTIATION

During gonadal development of zebrafish (*Danio rerio*), all individuals first develop ovaries regardless of their genotypic sex. In the presumptive males, which account for about half of the individuals, the oocytes then disappear from the gonads and this is followed by the appearance of spermatogonia and testicular differentiation (Takahashi, 1977). This phenomenon characterizes the undifferentiated type of gonads in the classification of Yamamoto (1969) and is also known as "juvenile hermaphroditism". Little is known about the mechanisms of juvenile sex reversal in zebrafish or in other gonochoristic showing the undifferentiated pattern of gonadal sex differentiation. The present study examined the possibility that apoptosis is involved in the disappearance of oocytes and sex reversal in zebrafish.

The fishes examined in this study were sampled from an all-female (XX chromosome type) group produced naturally by the mating of wild type females (XX females) with gynogenetic, sex-reversed females (XX males) and from mixed-sex groups produced by the mating of wild type females and males. The gynogenetic diploid fish were produced as described by Streisinger et al. (1981) and sex-reversed to phenotypic males by methyltestosterone administration. Samples were collected at 10, 15, 17, 19, 21, 23, 25, 27, 29, 31,

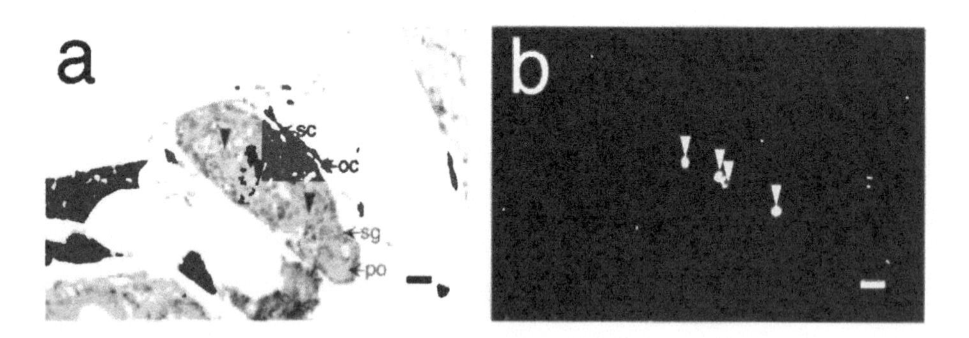

Fig. 5.3: Cross-section of the gonad of a 23-day-old zebrafish in the beginning of the sex change from ovary to testis. The section was stained consecutively by hematoxylin-eosin (a) and the TUNEL method (b). A large number of apoptotic oocytes were observed in fish between 23 and 25 days after hatching. TUNEL-positive cells (apoptotic oocytes) are indicated by black and white arrowheads in hematoxylin-eosin- and TUNEL-stained sections, respectively. oc, ovarian cavity; po, perinucleolar oocyte; sc, somatic cell; sg, spermatogonia. Bar represents 20 μm.

35, 40, 50 and 70 days after hatching and fixed in 10% neutral buffered formalin. Apoptosis of gonadal cells was examined in histological sections by in situ detection of fluorescein-labelled genomic DNA cleavage (TUNEL method). The presumptive sex of fish from the crosses between wild males and females was determined by histological examination of the gonads.

The gonads remained sexually indifferent until about 20 days after hatching. At 10 days after hatching, all gonads had a large number of oocytes up to the pachytene stage of development, but no apoptotic germ cells were found at this time. Small numbers of apoptotic oocytes were first detected in the gonads of all individuals at day 15 and their presence became conspicuous at day 19. Also, a flat ovarian cavity was formed in the gonads of all individuals by the fusion of cell outgrowths derived from the dorsal and ventral edges of the ovary between 14 and 20 days after hatching. A large number of apoptotic, early diplotene oocytes were observed in presumptive males between day 23 and day 25 (Fig. 5.3). Almost all apoptotic oocytes in these gonads disappeared by day 27. Apoptotic somatic cells were also found in the ovarian cavity of the presumptive males from 23 to 35 days after hatching. Moreover, apoptotic spermatocytes were detected at day 35, whereas apoptotic spermatogonia, spermatocytes, and spermatids at day 70 after hatching. In contrast, the females had a large number of apoptotic somatic cells in the connective tissue after day 23 but never presented any apoptotic cells associated with the ovarian cavity. Thus, the results of this study indicate a marked difference in the pattern of apoptosis of gonadal cells between males and females of zebrafish and suggest that apoptosis might be involved in the processes of testicular and ovarian differentiation. In particular, it seems plausible that the disappearance of apoptotic oocytes could be the trigger for testicular differentiation in males. To the best of our knowledge, this is the first example of apoptosis in association with the process of gonadal sex differentiation in fishes. Ongoing research is examining whether apoptosis is also involved in thermolabile sex determination using the pejerrey *O. bonariensis* as a model.

CONCLUDING REMARKS

The sex of teleost fishes is characterized by a high degree of plasticity and diverse forms of expression. Therefore, it is unlikely that a single event or gene could be used to explain the entire process of sex differentiation in one species, let alone different taxa. In this regard, the hypotheses presented in this study should be seen not as mutually exclusive but as complementary to each other and to the other hypotheses already put forth concerning the mechanism of sex differentiation in fishes. Nevertheless, mechanisms that are fundamental in one species may not be so in another. For instance, the accepted dependency on endogenous estrogens for the formation of the ovary, which has been assumed in many species (Adkins-Regan, 1987; Redding and Patiño, 1993; Patiño, 1997; Nakamura et al., 1998) and proven in some others (Guiguen et al., 1999; Kitano, 2000), has been recently challenged in medaka

(Kawahara and Yamashita, 2000). Thus, future studies must also address representative species from the various modes of sexuality.

As our understanding of the physiological and morphological phenomena associated with sex differentiation increases, we might be able to develop improved methods of sex control that are free of health hazards to the farmers, consumers, as also to the environment. For example, water temperature has been recently shown to affect sex differentiation in a number of species (Strüssmann and Patiño, 1995, 1999; Pandian and Koteeswaran, 1999), and this phenomenon has been seen as a "problem" since it may confound the effects of other treatments that are applied to control sex (Strüssmann and Patiño, 1995; Pandian et al., 1999). However, this knowledge could also be used to achieve synergistic effects, thus reducing the dependence on chemicals. In optimum scenarios, the use of temperature or other environmental factor could replace the current widespread use of chemicals for sex control in aquaculture (see Strüssmann et al. 1997). Also, knowledge of the endocrine and environmental control mechanisms of sex differentiation may be useful to predict the adverse effects of environmental changes such as global warming or pollution with endocrine disrupters on the reproductive fitness of feral fish populations. Needless is to say, feral fish populations are a major food resource that will become even more critical as the world's population continues to increase.

ACKNOWLEDGEMENTS

This study was supported by grants-in-aid from the Ministry of Education, Science, Sports and Culture of Japan (grants no. 12460085 and 15201003) and Tokyo University of Fisheries to CAS and from the United States Department of Agriculture (NRICGP 97-35203-4805) to RP. Anti–chum salmon ß-GTH1 and 2 were a kind donation from Dr H. Kawauchi (Kitasato University, Japan). We would like to thank Dr G. Yoshizaki and M. Kurita for help during the study.

REFERENCES

Adkins-Regan, E. 1987. Hormones and sexual differentiation. In: *Hormones and Reproduction in Fishes, Amphibians, and Reptiles* (eds), Norris, D.O. and Jones, R.E., Plenum Press, New York. pp. 1-29.

Baroiller, J.-F., Guiguen, Y. and Fostier, A. 1999. Endocrine and environmental aspects of sex differentiation in fish. Cell. Mol. Life Sci. 55: 910-931.

Bartley, D.M. and Hallerman, E.M. 1995. A global perspective on the utilization of genetically modified organisms in aquaculture and fisheries. Aquaculture 137: 1-7.

Carlton, J.T. 1992. Dispersal of living organisms into aquatic ecosystems as mediated by aquaculture and fisheries activities. In: Dispersal of living organisms into aquatic ecosystems (eds) Rosenfield, A. and Mann, R., University of Maryland Press, College Park, Maryland. pp. 13-48.

Crews, D. 1996. Temperature-dependent sex determination: The interplay of steroid hormones and temperature. Zool. Sci. 13: 1-13.

Crews, D., Bergeron, J.M., Bull, J.J., Flores, D., Tousignant, A., Skipper, J.K. and Wibbels, T. 1994. Temperature-dependent sex determination in reptiles: Proximate mechanisms, ultimate outcomes, and practical applications. Develop. Gen. 15: 297-312.

D'Cotta, H.D., Guiguen, Y., Govoroun, M., McMeel, O. and Baroiller, J.F. 1999. Aromatase gene expression in temperature-induced gonadal sex differentiation of tilapia *Oreochromis niloticus*. In: Proceedings of the Sixth International Symposium on the Reproductive Physiology of Fish (eds) Norberg, B., Kjesbu, O.S., Taranger, G.L., Andersson, E. and Stefansson, S.O., Fish Symp 99, Bergen, Norway. pp. 197-199.

De Silva, S.S. 1989. Exotic aquatic organisms in Asia. Asian Fisheries Society Special Publication No. 3, Asian Fisheries Society, Manila, Philippines. p. 154.

Feist, G. and Schreck, C.B. 1996. Brain-pituitary-gonadal axis during early development and sexual differentiation in the rainbow trout, *Oncorhynchus mykiss*. Gen. Comp. Endocrinol. 102: 394-409.

Fitzpatrick, M.S., Pereira, C.B. and Schreck, C.B. 1993. In vitro steroid secretion during early development of mono-sex rainbow trout: Sex differences, onset of pituitary control, and effects of dietary steroid treatment. Gen. Comp. Endocrinol. 91: 199-215.

Francis, R.C. 1992. Sexual lability in teleosts: developmental factors. Quart. Rev. Biol. 67: 1-18.

Guiguen, Y., Baroiller, J.-F., Ricordel, M. J., Iseki, K., McMeel, O.M., Martin, S.A.M. and Fostier, A. 1999. Involvement of estrogens in the process of sex differentiation in two fish species: The rainbow trout (*Oncorhynchus mykiss*) and a tilapia (*Oreochromis niloticus*). Mol. Reprod. Develop. 54: 154-162.

Hallerman, E.M. and Kapuscinski, A.R. 1995. Incorporating risk assessment and risk management into public policies on genetically modified finfish and shellfish. Aquaculture 137: 9-17.

Hawkins, M.B., Thornton, J.W., Crews, D., Skipper, J.K., Dotte, A. and Thomas, P. 2000. Identification of a third distinct estrogen receptor and reclassification of estrogen receptors in teleosts. Proc. Natl. Acad. Sci. USA . 97: 10751-10756.

Hunter, G.A. and Donaldson, E.M. 1983. Hormonal sex control and its application to fish culture. In: *Fish Physiology* Vol. IX, Part B Behaviour and Fertility Control (eds), Hoar, W.S., Randall, D.J. and Donaldson, E.M., Academic Press, New York. pp. 223-303.

Ito, L.S., Yamashita, M. and C.A. Strüssmann. 2003. Histological process and dynamics of germ cell degeneration in pejerrey *Odontesthes bonariensis* larvae and juveniles during exposure to warm water. J. Exp. Zool. 293: 492-499.

Kagawa, H., Kawazoe, I., Tanaka, H. and Okuzawa, K. 1998. Immunocytochemical identification of two distinct gonadotrophic cells (GTH I and GTH II) in the pituitary of bluefin tuna, *Thunnus thynnus*. Gen. Comp. Endocrinol. 110: 11-18.

Kawahara, T. and Yamashita, I. 2000. Estrogen-independent ovary formation in the medaka fish, *Oryzias latipes*. Zool. Sci. 17: 65-68.

Kitano, T., Takamune, K., Kobayashi, T., Nagahama, Y. and Abe, S. 1999. Suppression of P450 aromatase gene expression in sex-reversed males produced by genetically rearing female larvae at a high water temperature during a period of sex differentiation in the Japanese flounder (*Paralichthys olivaceus*). J. Mol. Endocrinol. 23: 167-176.

Kitano, T., Takamune, K., Nagahama, Y. and Abe, S. 2000. Aromatase inhibitor and 17a-methyltestosterone cause sex-reversal from genetical females to phenotypic males and suppression of P450 aromatase gene expression in Japanese flounder (*Paralichthys olivaceus*). Mol. Reprod. Develop. 56: 1-5.

Korpelainen, H. 1990. Sex ratios and conditions required for environmental sex determination in animals. Biol. Rev. 65: 147-184.

Leach, J.H. 1995. Non-indigenous species in the Great Lakes: Were colonization and damage to ecosystem health predictable? J. Aquat. Ecosyst. Health 4: 117-128.

Lin, Y.W., Rupnow, B.A., Price, D.A., Greenberg, R.M. and Wallace, R.A. 1992. *Fundulus heteroclitus* gonadotropins. 3. Cloning and sequencing of gonadotropic hormone (GtH) I and II β subunits using the polymerase chain reaction. Mol. Cell. Endocrinol. 85: 127-139.

Lodge, D.M., Stein, R.A., Brown, K.M., Covich, A.P., Brönmark, C., Garvey, J.E. and Klosiewski, S.P. 1998. Predicting impact of freshwater exotic species on native biodiversity: Challenges in spatial scaling. Aust. J. Ecol. 23: 53-67.

Mal, A.O., Swanson, P. and Dickhoff, W.W. 1989. Immunocytochemistry of the developing salmon pituitary gland. Am. Zool. 29: 94A.

Miranda, L.A., Strüssmann, C.A. and Somoza, G.M. 2001. Immunocytochemical identification of GtH1 and GtH2 cells during the temperature-sensitive period for sex determination in pejerrey, *Odontesthes bonariensis*. Gen. Comp. Endocrinol., 124: 45-52.

Miranda, L.A., Strobl-Mazzulla, P.H., Strüssmann, C.A., Parhar, I. and Somoza, G.M. 2003. Gonadotropin-releasing hormone neuronal development during the sensitive period of temperature sex determination in the pejerrey fish, *Odontesthes bonariensis*. Gen. Comp. Endocrinol., 132: 444-453.

Moyle, P.B. 1991. Ecology of introduced fishes and aquatic invertebrates in North America. In : Ecology and Management of the Zebra Mussel and Other Introduced Aquatic Nuisance Species, Introduced Species Workshop (ed.), Yount, J.D., United States Environmental Protection Agency, Washington. pp. 1-3.

Nakamura, M., Kobayashi, T., Chang, X-T. and Nagahama, Y. 1998. Gonadal sex differentiation in teleost fish. J. Exp. Zool. 281: 362-372.

Nozaki, M., Naito, N., Swanson, P., Miyata, K., Nakai, Y., Oota, Y., Suzuki, K. and Kawauchi H. 1990a. Salmonid pituitary gonadotrops. I. Distinct cellular distributions of two gonadotropins GTHI and GTHII. Gen. Comp. Endocrinol. 77: 348-357.

Nozaki, M., Naito, N., Swanson, P., Dickhoff, W., Nakai, Y., Suzuki, K. and Kawauchi, H.1990b. Salmonid pituitary gonadotrops. II. Ontogeny of GtHI and GtHII cells in the rainbow trout *(Salmo gairdneri irideus)*. Gen. Comp. Endocrinol. 77: 358-367.

Pakdel, F., Le Guellec, C., Vaillant, C., Le Roux, M.G. and Valotaire, Y. 1989. Identification and estrogen induction of two estrogen receptors (ER) messenger ribonucleic acids in the rainbow trout liver: Sequence homology with other ERs. Mol. Endocrinol. 3: 44-51.

Pandian, T.J. and Koteeswaran, R. 1999. Lability of sex differentiation in fish. Curr. Sci. 76: 580-583.

Pandian, T.J., Venugopal, T. and Koteeswaran, R. 1999. Problems and prospects of hormone, chromosome and gene manipulations in fish. Curr. Sci. 76: 369-386.

Patiño, R. 1997. Manipulations of the reproductive system of fishes by means of exogenous chemicals. Prog. Fish-Cult. 59: 118-128.

Patiño, R. Davis, K.B., Schoore, J.E., Uguz, C., Strüssmann, C.A., Parker, N.C., Simco, B.A. and Goudie, C.A. 1996. Sex differentiation of channel catfish gonads: Normal development and effects of temperature. J. Exp. Zool. 276: 209-218.

Patiño, R., Xia, Z., Gale, W.L., Wu, C., Maule, A.G. and Chang, X. 2000. Novel transcripts of the estrogen receptor α gene in channel catfish. Gen. Comp. Endocrinol. 120: 314-325.

Pieau, C. 1996. Temperature variation and sex determination in reptiles. BioEssays, 18: 19-26.

Redding, J.M. and Patiño, R. 1993. Reproductive physiology. In: The Physiology of Fishes (ed.), Evans, D. HCRC Press, Boca Raton. pp. 503-534.

Salbert, G., Bonnec, G., Le Goff, P., Boujard, D., Valotaire, Y. and Jego, P. 1991. Localization of the estradiol receptor mRNA in the forebrain of rainbow trout. Mol. Cell. Endocrinol. 76: 173-180.

Streisinger, G., Walker, C., Dower, N., Knauber, D. and Singer, F. 1981. Production of clones of homozygous diploid zebra fish (*Brachydanio rerio*). Nature 291: 293-296.

Strüssmann, C.A. and Nakamura, M. 2002. Morphology, endocrinology, and environmental modulation of gonadal sex differentiation in teleost fishes. Fish Physiol. Biochem. 26: 13-29.

Strüssmann, C.A. and Patiño, R. 1995. Temperature manipulation of sex differentiation in fish. In: Proceedings of the Fifth International Symposium on the Reproductive Physiology of Fish (eds.), Goetz, F.W. and Thomas, P. Fish Symp 95, Austin, Texas. pp. 153-160.

Strüssmann, C.A. and Patiño, R. 1999. Sex Determination, Environmental. In: Encyclopedia of Reproduction (eds), Knobil, E. and Neill, J.D. Vol. 4,. Academic Press, New York. pp. 402-409

Strüssmann, C.A., Moriyama, S., Hanke, E.F., Calsina Cota, J.C. and Takashima, F. 1996b. Evidence of thermolabile sex determination in pejerrey. J. Fish Biol. 48: 643-651

Strüssmann, C.A., Saito, T. and Takashima, F. 1998. Heat-induced germ cell deficiency in the teleosts *Odontesthes bonariensis* and *Patagonina hatcheri*. Comp. Biochem. Physiol. 119A: 637-644.

Strüssmann, C.A., Saito, T., Usui, M., Yamada, H. and Takashima, F. 1997. Thermal thresholds and critical period of thermolabile sex determination in two atherinid fishes, *Odontesthes bonariensis* and *Patagonina hatcheri*. J. Exp. Zool. 278: 167-177.

Strüssmann, C.A., Takashima, F. and Toda, K. 1996a. Sex differentiation and hormonal feminization in pejerrey *Odontesthes bonariensis*. Aquaculture 139: 31-45.

Takahashi, H. 1977. Juvenile hermaphroditism in the zebrafish, *Brachydanio rerio*. Bull. Fac. Fish. Hokkaido Univ. 28: 57-65.

Thorgaard, G.H. 1983. Chromosome set manipulation and sex control in fish. In: *Fish Physiology* Vol. IX, Part B Behavior and Fertility Control (eds), Hoar, W.S., Randall, D.J. and Donaldson, E.M., Academic Press, New York. pp. 405-434.

Toubeau, G., Poilve, A., Baras, E., Nonclercq, D., De Moor, S., Beckers, J.F., Dessy-Doize, C. and Heuson-Stiennon, L. 1991. Immunocytochemical study of cell type distribution in the pituitary of *Barbus barbus* (Teleostei, Cyprinidae). Gen. Comp. Endocrinol. 83: 35-47.

Van den Hurk, R. 1982. Effects of steroids on gonadotropic (GTH) cells in the pituitary of rainbow trout, *Salmo gairdneri*, shortly after hatching. Cell Tiss. Res. 224: 361-368.

Van Winkoop, A., Timmermans, L.P.M. and Goos, H.J.Th. 1994. Stimulation of gonadal and germ cell development in larval and juvenile carp (*Cyprinus carpio* L.) by homologous pituitary extract. Fish Physiol. Biochem. 13: 161-171.

Xia, Z., Patiño, R., Gale, W.L., Maule A.G. and Densmore, L.D. 1999. Cloning, in vitro expression, and novel phylogenetic classification of a channel catfish estrogen receptor. Gen. Comp. Endocrinol. 113: 360-368.

Xia, Z., Gale, W.L., Chang, X., Langenau, D., Patiño, R., Maule A.G. and Densmore, L.D. 2000. Phylogenetic sequence analysis, recombinant expression, and tissue distribution of a channel catfish estrogen receptor beta. Gen. Comp. Endocrinol. 118: 139-149.

Yamamoto, T. 1969. Sex differentiation. In: Fish Physiology Vol. III, Reproduction and Growth, Bioluminescence, Pigments, and Poisons (eds), Hoar, W.S. and Randall, D.J., Academic Press, New York. pp. 117-175.

Yoshiura, M., Suetake, H. and Aida, K. 1999. Duality of gonadotropin in a primitive teleost, Japanese eel (*Anguilla japonica*). Gen. Comp. Endocrinol. 114: 121-131.

Chapter 6

Genes for Fish GnRHs and Their Receptor: Relevance to Aquaculture Biotechnology

D. Alok* and **Y. Zohar**
Center of Marine Biotechnology, University of Maryland Biotechnology Institute,
701E. Pratt Street, Baltimore MD 21202, USA.
email: alok.deoraj@sygeninternational.com.

ABSTRACT

The sustainability of the commercial aquaculture undoubtedly depends on the continuous and predictable supply of quality seeds of the cultivated species. Propagation of fish culture has been impeded by the inadequate supply of fish seed—in terms of both quality and quantity. Efforts have been made to establish captive broodstock and induce their spawning. However, it was demonstrated that in captivity, a majority of farmed fish species fail to reproduce successfully, primarily due to the absence of the release of luteinizing hormone (LH) into the circulation. LH is released after the binding of gonadotropin-releasing hormone (GnRH) to its specific receptor(s) present on the gonadotrophic cells of the pituitary gland. Current spawning induction methods use non-endogenous GnRH agonists. Three different endogenous forms of GnRH (chicken GnRH-II, salmon GnRH and seabream GnRH) were discovered in the brain of perciform species. It was further demonstrated that only one form of the GnRH, seabream GnRH (sbGnRH) reaches the pituitary in abundance. The correlation of sbGnRH levels with gonadal development and its specific anatomical delivery system for sbGnRH suggest that this form is the most relevant endogenous form that induces LH release, final oocyte maturation and spawning. However, sbGnRH elicited a very low level of LH release compared to the chicken (c)

* Present address : SyGen International, 3033 Nashville Road FREANKLIN, KY 42134 USA

GnRH-II form in *in vivo* studies carried out in seabream and striped bass. To understand the molecular mechanisms controlling reproduction, cDNAs and genes for all three forms of GnRHs have been isolated from perciform species. Very recently, a full-length cDNA for the pituitary GnRH receptor was cloned from striped bass. A major portion of pituitary GnRH-R cDNA was also cloned from gilthead seabream. This chapter discusses current spawning methods in the light of possible isoforms of the GnRH receptor with reference to the presence of multiple GnRHs. It concludes with a summary of the relevance of molecular interaction of GnRH-GnRH receptor(s) in modern aquaculture biotechnology.

Keywords: Aquaculture, biotechnology, fish, gonadotropin-releasing hormone (GnRH), genes, neuroendocrine, pituitary, reproduction, signal transduction, spawning.

INTRODUCTION

Finfish culture in a controlled environment is one of the primary components of the aquaculture industry. The global increase in seafood demand, persistent threats of pollution to our natural resources and diminishing returns from capture fisheries in the recent years have attracted attention from fish farmers and scientific community to promote aquaculture of important finfish species. To sustain the finfish industry, a need for diversification, conservation and stock enhancement of identified species has also been realized. There is no doubt that larval rearing, nutrition, disease resistance and specific niche have profound influence on the propagation of any species, but a predictable seed supply is the key to a successful program for finfish aquaculture.

Finfish are one of the most diverse groups of vertebrates. Identification of aquaculturable species, their stock enhancement and ensuring their uninterrupted seed supply offer challenges to the fish growers and scientific community. Therefore, it is essential to elucidate the reproductive mechanisms that regulate seed production or spawning at both the physiological and molecular levels. Recent advancements to our knowledge of physiological and environmental controls of fish reproduction have yielded biotechnological tools for a substantial control over the predictablity of the seed supply (Zohar and Mylonas, 2001). This chapter briefly reviews the recent advacnes in molecular mechanism controlling reproduction in finfish.

CONTROL OF REPRODUCTION

The brain-pituitary-gonadal (BPG) web (Fig. 6.1) controls reproduction in teleost species. Different centers of brain receive environmental (e.g. photoperiod, temperature, pH, salinity), social and nutritional cues to trigger the release of a consortium of neuropeptides, neurotransmitters and neurosteroids. These factors, via their positive and negative feedback mechanisms, regulate spatiotemporal synthesis and release of the gonadotropin-releasing hormone (GnRH).

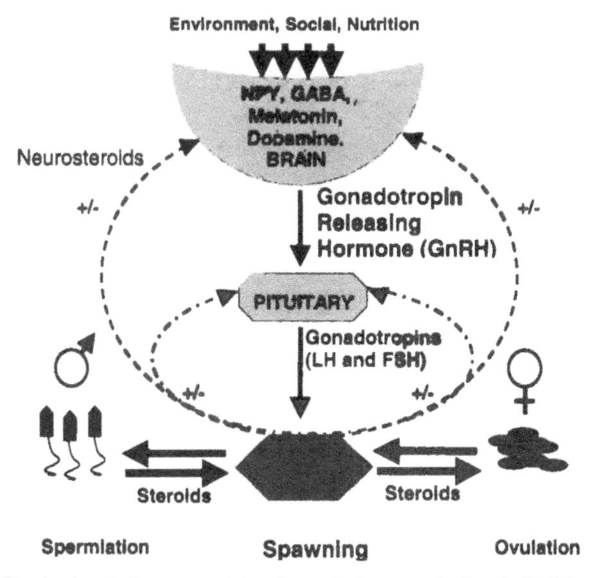

Fig 6.1: The brain-pituitary-gonadal web controls reproduction in finfish species. This web is accessible to intervention at the brain, pituitary and gonadal level for induced spawning by environmental, GnRH, pituitary extract or human chorionic gonadotropin (hCG) treatments.

GnRH is a central molecule of the neuroendocrine web, which, upon release, binds to the GnRH-receptor (GnRH-R) present on the gonadotrophic cells in the pituitary so as to initiate an intracellular signaling cascade of synthetic and release pathways of gonadotropins (GtHs). Gonadotropins, luteinizing hormone (LH) and follicle-stimulating hormone (FSH) act on the gonads to initiate steroidogenesis and gametogenic processes, resulting in final oocyte maturation (FOM), spermiation, ovulation and spawning. Although the basic mechanisms of teleost reproductive processes remain the same, there is enormous variation in their reproductive behavior and seasonality.

CAPTIVITY-INDUCED REPRODUCTIVE DYSFUNCTION

Commercial fish seed production still relies on the unpredictable acquisition of the limited wild broodstock undergoing FOM and on their induced spawning. While developing reproductive technologies for finfish spawning, a captivity-induced disruption in the completion of oocyte maturation and spermiation was observed in many farmed fish. In males, the reproductive disruption results into smaller volume of milt, i.e. lower quality of milt (Zohar and Mylonas, 2001). In the case of females, the reproductive disruption is exemplified at three levels: (1) complete absence of vitellogenesis, e.g. in *Anguilla* and *Seriola*; (2) arrest of FOM, e.g. in *Morone saxatilis*; and (3) lack of spawning, e.g. in salmon (Bromage et al., 1992), catfish (Alok et al., 1993) and grouper (Hassin et al., 1997). In some cases, manipulations of temperature, photoperiod, vegetation, tank depth and salinity have yielded predictable spawning. However, in majority of farmed fish, hormonal administration is the only solution to induce reliable spawning (Zohar and Mylonas, 2001). Interestingly, with prolonged maintenance in captivity, fish species can be domesticated and their spawning performance improved (e.g. gilthead seabream: Zohar, 1989 and grouper: Hassin et al. 1997). Since the discovery of hypophyzation methods in the early 1930s, administration of homologous and heterologous pituitary extracts (PE) and subsequently hCG in the 1970s, became popular among fish farmers to correct the failure of captive broodstock to spawn spontaneously (Hodson and Sullivan, 1993; Zohar et al., 1989b). This also indicated that the primary reason for the reproductive failure in most of the farmed fish is possibly due to the hormonal imbalance that prevents gametogenesis and spawning. For example, in striped bass approximately 85-90% of fish fail to initiate FOM in captivity (Woods and Sullivan, 1993; Blythe et al., 1994; Mylonas et al. 1998). In general, oocyte maturation is arrested at the end of vitellogenesis (Steven, 1999). Studies in striped bass, determined that pituitary LH is synthesized but not released into the blood of 80% of the wild fish acquired for hatcheries purpose, nor in any of the captive reared fish (Mylonas et al., 1998). This suggested that most of farmed fish do not complete gonadal development. Moreover, their oocytes undergo rapid atresia during the spawning season due to the lack of LH release from the pituitary (Mylonas et al., 1997a,c,e, 1998; Steven, 1999; Zohar 1996; Zohar and Mylonas, 2001). During the last couple of decades, research has been focused on the development of GnRH-based spawning induction methods and elucidation of the mechanism of GnRH-GnRH-R interactions leading to LH synthesis and release.

GnRH-BASED SPAWNING TECHNOLOGY

During the early 1970s, the amino acid sequence of the mammalian (m) luteinizing hormone releasing hormone (LHRH) was deduced. The hypothalmic LHRH causes the release of LH and FSH from the pituitary (Matsuo et al., 1971; Burgus et al., 1972). Later, a similar molecule with LH releasing potency was identified from salmon brain (Sherwood et al., 1983). This observation indicates that GnRH has been conserved during evolution (Table 6.1). GnRH is a decapeptide which acts at higher levels in the BPG web, presumably eliciting an integrated responses for coordinated induction of LH release, ovulation and spawning in fish (De Leeuw et al., 1985; Lin et al., 1988; Crim et al., 1988; Zohar et al., 1989a; Nandeesha et al., 1990; Fermin, 1991). However, the native forms of GnRHs are very rapidly cleaved at positions 6 and 9 by specific enzymes present in various fish tissues and cleared quickly (5-7 min.) from the fish circulation (Goren et al., 1990; Gothilf and Zohar, 1991). Conversely, GnRH-agonists (GnRHa), designed by replacing D-amino acid at position 6 and by modifying position 9, are resistant to enzymatic degradation (Zohar et al., 1990b;

Alok et al., 1999). They are cleared slowly from the fish's circulation (Gothilf and Zohar, 1991) and are very potent in inducing sustained LH release for periods lasting up to 72 hr, depending on fish species and water temperature. Since then, highly potent mGnRHa and sGnRHa (Zohar and Mylonas, 2001) have gained worldwide acceptance to induce FOM and spawning in a variety of sexually mature fish (Zohar, 1989, 1996; Breton et al., 1990; Alok et al., 1993, 1995 and 1997; Zohar and Mylonas, 2001).

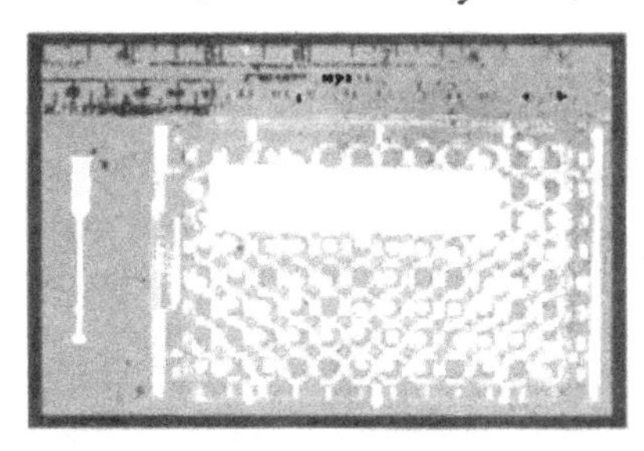
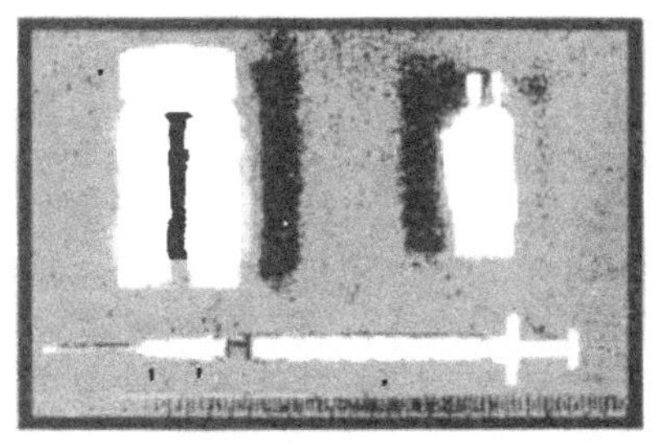

Fig 6.2: Ethylene-vinyl-acetate polymer mGnRHa implants and p[Fad-sa] microspheres. GnRH implants are manufactured with a fixed dosage of GnRHa but in microspheres, the volume can adjust the administered dosage. By changing the ratio of bulking agents, controlled delivery of GnRHa in fish blood can be extended upto 30 days while microspheres can release GnRHa for upto 60 days

However, acute administration of even the most potent GnRHa was not sufficient to induce the long-term elevations in plasma LH required for the initiation and completion of FOM and ovulation in some fishes, including striped bass (Zohar, 1996; Mylonas and Zohar, 1998; Zohar and Mylonas, 2001). Therefore, a number of GnRHa delivery systems have been tested in fish for sustained release over the last couple of decades. These tests included cholesterol implants, in which GnRHa release rates in the fish circulation are adjusted by changing the percentage of cholesterol and cellulose in the matrix. However, highly variable GnRHa release rates are one of the primary negative features of this delivery system (Carolsfeld et al. 1988). In the early 1990s, biodegradable microspheres made of co-polymers of lactic acid and glycolic acid (LGA) were tested for induced spawning. GnRHa release rates were controlled by changing the ratio of lactic acid to glycolic acid and its molecular weight (Zohar, 1988; Zohar et al., 1990a Breton et al., 1990; Chasin and Langler, 1990). Recently, a biodegradable microspheres made of polyanhydride co-polymers of fatty acid dimer and sebasic acid (Fad-sa) and implants of non-biodegradable Ethylene-vinyl-acetate (E-Vac) polymers have been developed for sustained delivery of [D-Ala^6, Pro^9 NEt]-mGnRH (Mylonas et al., 1995a; Zohar, 1996; Fig 6.2). Fad-sa microspheres are prepared using the double emulsion and solvent evaporation process which can release GnRHa for at least 8 weeks. Administration of microspheres is convenient and practical because the same preparation can be administered in a suspension vehicle to treat fish of different sizes from a few grams to several kgs. On the other hand, EVac implants are manufactured in the from of 2mm monolithic discs and administered intramuscularly, using an implanter. In the EVac implants, the GnRHa is mixed with an inert bulking agent and then entrapped in the EVac matrix. GnRHa release rates are adjusted by modifying the molecular weight of the polymer and by changing the percentage of the bulk loading agent in the matrix. These implants and microspheres have been employed successfully to induce FOM, ovulation and spawning in a variety of commercially important fish (Zohar et al. 1995b, 1996; Mylonas et al., 1995a,b, 1996, 1997a,b, c,d,e, 1998; Mylonas and Zohar, 1998). EVac GnRHa implants have a long shelf life and can be effective for upto 3 years . Properly timed GnRHa-EVac implant administration can sustain an elevated plasma LH levels for periods lasting upto 2-5 weeks.

Some fish, e.g. goldfish, carp, (Peter et al., 1987) and catfish, (Alok et al., 1995, 1997) have strong dopaminergic inhibitory control over LH release and, therefore, administration of mGnRHa or sGnRHa has to be combined with a dopamine antagonist for LH release, spawning, and ovulation. In some studies, mGnRHa elicited LH release or induced spawning, only when used in combination with a dopamine receptor antagonist, viz., pimozide or domperidone (Kaul and Rishi 1986; Lin et al., 1986; Ramos, 1986; Fermin, 1991; Glubokov et al., 1991; Tharakan and Joy 1996). sGnRHa, alone or in combination with dopamine antagonists, have also been shown to induce spawning in a variety of species such as carp and catfishes (Nandeesha et al., 1990; Alok et al., 1993, 1995, 1997). Currently, acute or sustained administration of GnRHas for LH release, ovulation and spawning in fish has become a method of choice in recent years to correct captivity-induced reproductive dysfunction in farmed fish (see review, Zohar and Mylonas, 2001).

For the last two decades, the role of GnRH in controlling vertebrate reproduction has been an area of intense research. Altogether, there are 13 forms of endogenous GnRHs that have been identified and characterized from vertebrate species (Table 6.1). GnRH is highly conserved in length and in amino acid residues at positions 1, 2, 4, 9 and 10. Position 8 is the most variable, followed by positions 5, 7 and 6 (King and Millar, 1995). Each vertebrate species has at least two forms of GnRH with chicken (c) GnRH-II ubiquitously found in all the vertebrate species, including mammals (Ngamvonchon et al., 1992a; Lescheid et al., 1997; Sherwood et al., 1997). In the mid 1990s, studies on perciform fish identified three endogenous forms of GnRH (seabream (sb) GnRH, cGnRH II and salmon(s) GnRH) in the brains of gilthead seabream (*Sparus aurata*; Powell et al., 1994) and striped bass (*Morone saxatilis*; Gothilf et al., 1995a).

Subsequently, three GnRH forms have been found in other perciform species and fish of different orders (Powell et al., 1995, 1997; Weber et al., 1997; Okubo et al., 2000a). The presence of a third form of GnRH has also been indicated by immunohistochemistry, in human, bovine and rat brains (Yahalom et al., 1999). In seabream and striped bass, all three endogenous GnRHs are potent LH secretogouges (Zohar et al., 1995a; Blaise et al., 1996).

Recently, genes for individual forms of endogenous GnRHs have been isolated in several fish species (White and Fernald; 1998; Coe et al., 1995). In striped bass, we have isolated three separate genes encoding for sbGnRH, cGnRH II and sGnRH (Chow et al. 1998). Structural analyses of genes and

Table 6.1: Amino acid sequences of 13 known vertebrate GnRH forms are shown along with the conventionally accepted nomenclature. GnRH forms are named after the animal in which they are first identified. Guinea pig (gp), mammalian (m), chicken (c), seabream (sb), pejerrey (p), herring (hr), catfish (cf), frog (r), salmon (s), dogfish (df) and lamprey (l). GnRHs are listed in order of similarity to mGnRH. Amino acid differences are highlighted with shaded background.

Endogenous GnRHs, neurotransmitters and neuropeptides

	1	2	3	4	5	6	7	8	9	10
GpGnRH	pGlu	[illegible]	Trp	Ser	Tyr	Gly	Val	Arg	Pro	Gly-NH_2
mGnRH	pGlu	His	Trp	Ser	Tyr	Gly	Leu	Arg	Pro	Gly-NH_2
cGnRH-I	pGlu	His	Trp	Ser	Tyr	Gly	Leu	Gln	Pro	Gly-NH_2
SbGnRH	pGlu	His	Trp	Ser	Tyr	Gly	Leu	Ser	Pro	Gly-NH_2
PGnRH	pGlu	His	Trp	Ser	Phe	Gly	Leu	Ser	Pro	Gly-NH_2
HrGnRH	pGlu	His	Trp	Ser	His	Gly	Leu	Ser	Pro	Gly-NH_2
CfGnRH	pGlu	His	Trp	Ser	His	Gly	Leu	Asn	Pro	Gly-NH_2
RGnRH	pGlu	His	Trp	Ser	Tyr	Gly	Leu	Trp	Pro	Gly-NH_2
sGnRH	pGlu	His	Trp	Ser	Tyr	Gly	Trp	Leu	Pro	Gly-NH_2
cGnRH-II	pGlu	His	Trp	Ser	His	Gly	Trp	Tyr	Pro	Gly-NH_2
dfGnRH	pGlu	His	Trp	Ser	His	Gly	Trp	Leu	Pro	Gly-NH_2
lGnRH-III	pGlu	His	Trp	Ser	His	Asp	Trp	Lys	Pro	Gly-NH_2
lGnRH-I	pGlu	His	Tyr	Ser	Leu	Glu	Trp	Lys	Pro	Gly-NH_2

promoter regions have revealed key information to link the expression of fish GnRHs in response to changing environmental, social and nutritional milieu (Fig. 6.3). The putative upstream AP-1, Glucocorticoid Response Elements (GRE), Estrogen Response Elements (ERE), SF-1 domains and multiple start sites in the sbGnRH promoter region, Pit-1, Oct-1, AP-1 domains in cGnRH II promoter, Sox sites and multiple start sites in the sGnRH promoter (Coe et al., 1995) suggest their spatiotemporal expressions. The information on the gene structure has clearly implicated all three GnRHs for their role in integrating

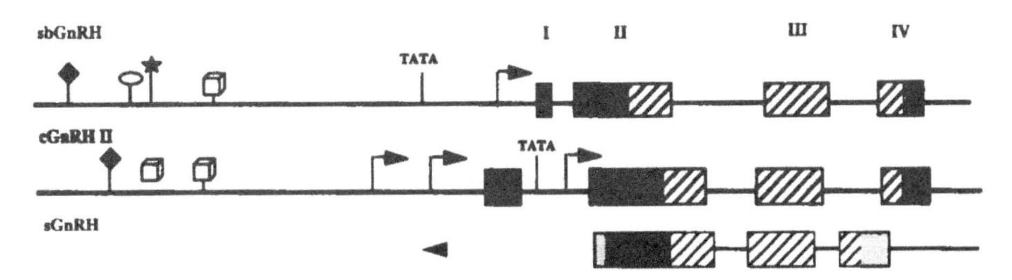

Fig 6.3: Three GnRH genes isolated from striped bass. Exons are numbered I, II, III and IV. Dotted arrow indicates undetermined sequence. Exon designations: = 5' untranslated region; ■ = GnRH decapeptide; = GAP (GnRH associated peptide); Arrows indicate putative transcriptional start sites. = ERELMs (estrogen response-like elements); =GRE (glucocorticoid response element); ◆= AP-1 site; and ★= SF-1 site (steroidogenic factor-1).

the physiological responses, which presumably provide a more balanced stimulation to the reproductive events culminating in FOM, ovulation, spermiation and spawning. Isolation of separate cDNAs has been another hallmark of fish GnRH research. Several fish-specific GnRH cDNAs have been isolated recently that demonstrated a typical blueprint of their structural organization (Fig 6.4). cDNAs of endogenous GnRHs consist of 5' untranslated region (UTR) of different length, a signal peptide region, conserved decapeptide region, cleavage site, a highly variable GnRH associated peptide (GAP) and 3' UTR (Gothilf et al. 1996).

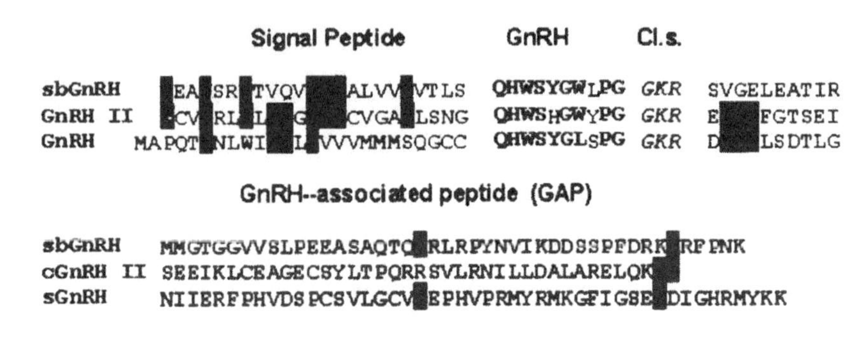

Fig 6.4: Deduced amino acid sequences from seabream cDNAs of endogenous GnRH precursors exhibiting regions of signal peptide, GnRH decapeptide (shown in bold letters), conserved cleavage site (Cl.s., shown in italics) and the highly variable GAP region. Shaded area represents the homologous region (Gothilf et al., 1996).

In contrast to the mammalian system, fish lack a functional portal system for the delivery of GnRHs in the pituitary. In fish, the hypothalamo-pituitary communication is by direct neuronal innervation (Kah et al. 1993). Since the discovery of the three forms of GnRH in the brain, seasonal profiles for each form have been resolved by ribonuclease protection assays in seabream (Gothilf et al., 1997) and by specific ELISAs in seabream and striped bass (Holland et al., 1998, 2001) in relation to the gonadal development, FOM, ovulation and spawning. By using ribonuclease protection assay (RPA) and high pressure liquid chromatography (HPLC), it was determined that there is a differential level of synthesis of all three GnRHs, with sbGnRH being the most abundant form present in the brain of a sexually mature fish (Powell et al., 1994; Gothilf et al., 1996). The dominance and distribution of species-specific forms of GnRHs in the pituitaries and specific neuronal projections for their delivery has also been shown in different fish species (Montero, et al., 1994; Zandbergen et al., 1995; Parhar and Sukuma, 1997; Okuzawa et al., 1997; Yamamoto et al., 1998). In general, sbGnRH levels are 10-20 fold higher than cGnRH II and 100-200 fold higher than sGnRH. Similar ratios of the mRNA levels of the three GnRH forms were found in wild (FOM stage) and captive striped bass (post-vitellogenic, arrested FOM stage). However, ELISA results indicate significantly higher levels of sbGnRH and sGnRH in the pituitary of captive fish at the early

FOM stage (Steven et al., 1999) when compared to the wild fish. This observation implies that it is unlikely that a deficient delivery of any of the endogenous GnRHs to pituitary may be a cause of reproductive dysfunction.

Although GnRH is the primary regulator of FSH and LH synthesis and release, other neuropeptides, neurotransmitters and steroids found in the hypothalamus, pituitary and gonads also influence and possibly regulate GnRH action. Norepinephrine stimulates GnRH release from the hypothalamus (but not pituitary), whereas serotonin stimulates and dopamine inhibits its release from both sites *in vitro* (Chang et al. 1996). GnRH-mediated release of LH from dispersed pituitary cells is inhibited by dopamine but stimulated by norepinephrine, serotonin, vasotocin, and neuropeptide Y (Groves and Batten, 1986; Peter et al., 1990; Somoza and Peter, 1991). Melatonin inhibits GnRH-induced release of LH and FSH in the neonatal rat, but not in the adult rat. Recently, a fine control mechanism on the GnRH function has been suggested that is mediated by neurosteroids via a GABA-ergenic pathway (Gennazani et al., 2000). In mammals, it has also been shown that the pattern of co-expression of galanin and GnRH genes in hypothalamic neurons is sexually dimorphic, with much greater galanin mRNA content in the GnRH neurons of females as compared to males. Such multihormonal control of GnRH action on LH synthesis and release undoubtedly allows for very precise control over GnRH-GnRH-R interactions.

GONADOTROPIN RELEASING HORMONE RECEPTOR

In vertebrates, the pituitary GnRH-R plays a pivotal role in determining the pituitary's responsiveness to GnRH-binding and, therefore, controls LH and FSH synthesis and release. It has been demonstrated that captivity did not influence the differential levels of the three GnRH forms delivered to the pituitary and also that all three GnRHs induced LH release when administered exogenously with cGnRH-II exhibiting maximum potency, followed by sGnRH and sbGnRH (Zohar et al., 1995a; Blaise et al., 1996). These results have raised important questions regarding the type of GnRH-R present on the gonadotrophic pituitary cells and their differential response to the closely related three GnRH forms. Alterations in the number and affinity of pituitary GnRH-Rs associated with seasonal variations and regulation of LH release have been reported in goldfish (Omeljaniuk et al., 1989; Habibi et al., 1989). Some studies in mammals have documented no change in the pituitary receptor mRNA levels (Mason et al., 1994) at different stages of gonadal development. But, in general, increases in the mRNA levels have been correlated with pituitary GnRH-R number and its sensitivity to GnRHs (Bauer-Dantoin et al., 1993; Brooks et al., 1993; Turzillo et al., 1994; Kakar et al., 1994). Padmanabhan et al., 1995; In striped bass, a significant increase in pituitary GnRH-R mRNA levels was observed in fish that showed higher gonado-somatic index (GSI= Gonad weight/body weight × 100) values when compared to fish with lower GSI values (Alok et al., 2000). Seasonal variation or captivity-induced changes in the GnRH-R mRNA levels are also expected in the pituitaries of farmed fish. Changes in the GnRH-R mRNA level may also correspond with LH concentration in the pituitary and oocyte diameters (Alok et al. 2000). Monitoring the rise in the GnRH-R mRNA levels in the pituitaries may pinpoint the precise physiologically compatible stage at spawning.

Studies of fish GnRH-R from pituitary membrane preparations have been limited to catfish, winter flounder, goldfish and seabream (De Leeuw et al., 1988; Crim et al., 1988; Habibi et al., 1987; Pagelson and Zohar, 1992). In recent years, numerous fish GnRH-receptor cDNAs have been cloned and characterized (Tensen et al., 1997; Illing et al. 1999, Alok et al., 2000; Okubo et al., 2000b; Madigou et al., 2000; Robison et al., 2001). The GnRH-R gene has been isolated primarily from mammalian species (human, mouse, rat, sheep and dog), which contains three exons (Sealfon et al., 1997). The regulatory sequence in 5' flanking region of human GnRH-R gene includes a putative cyclic AMP (cAMP) response elements and glucocorticoid/ progesterone response elements (GRE/PRE). It also includes PEA-3 (a phorbol ester response element) and an AP-2 recognition site. Most interestingly, it included Pit-1, an anterior pituitary specific transcriptional factor. It is notable that thus far, only one form of GnRH-R cDNAs has been isolated from the pituitaries of all mammalian and non-mammalian vertebrate species, including striped bass (Alok et al., 2000). Nonetheless, discovery of multiple GnRH-R from pituitaries will point to their specificity to the multiple GnRHs present in the species. Delineating their annual profile in farmed fish will help determine the treatment regimen with compatible and physiologically relevant endogenous GnRHs. On the other hand, the activity of a single form of GnRH-R present on the gonadotrophic cells may be specifically triggered, depending on the differential amounts of individual GnRH forms present in the pituitary (Alok et al., 2001). In addition to the pituitary GnRH-Rs cloning, discovery of GnRH-R from goldfish brain (Illing et al., 1999), two different GnRH-R forms (Wang et al., 2001) from bullfrog brain and type II GnRH-R from primates (Neil et al., 2001) and mormoset (Millar et al., 2001) have further added a level of complexity to the GnRH-GnRH-R interactions controlling reproduction in vertebrates (Stojlkovic et al., 1994; Sealfon et al., 1997).

MOLECULAR BASIS OF GnRH-GnRH-R INTERACTIONS

All GnRH-Rs isolated from non-mammalian vertebrates, including fish, contain a C-terminal tail. It was shown that only those GnRH-Rs having C-terminal tails show desensitization of phosphoinositide responses within 5-10 min. of agonist challenge. Type I mammalian GnRH-R fails to undergo agonist-dependent phosphorylation due to the lack of intracellular C-terminal tail that results directly in a resistance to rapid desensitization (Lin et al., 1998; Willars et al., 1999). Pawson et al., (1998) have demonstrated that internalization of chicken GnRH-R is at least 12-fold faster than human GnRH-R in COS-1 mammalian cells. Therefore, the process of internalization, downregulation and desensitization has to be looked again in conjunction with the functional levels of the endogenous GnRHs in the pituitaries of fish species.

```
1   MNTTLCDSAV AMYHLTTDHQ LNASCNYSSP TSNWTSGGGS

41  LQLPTFTTAA KVRVIITCIL CGISAFCNLA VLWAAHSDGK
               I          TMD
81  RKSHVRVLII NLTVADLLVT FIVMPVDAVW NITVQWLAGD
              II         TMD
121 LACRLLMFLK LQAMYSCAFV TVVISLDRQS AILNPLAINK
     III        TMD
161 ARKRNRVMLT VAWGMSVVLS VPQLFLFHNV TIIYPEDFTQ
              IV         TMD
201 CTTRGSFVTH WHETAYNMFT FSCLFLLPLI IMITCYTRIF
                               V         TMD
241 CEISKRLKKD NLPSNEVHLR RSKNNIPRAR MRTLKMSIVI
                                         VI
281 VSSFIVCWTP YYLLGLWYWF FPDDLEGKVS HSLTHILFIF
          TMD                             VII
321 GLVNACLDPV IYGLFTIHFR KGLRRYYCNA TKASDLDNNT
          TMD
361 VITGSFICAA NSLPLKREWC QPGEPVLYSD NHSRAELTSP
401 RSSFLRDPNQ SSPESNL
```

Fig 6.5: Amino acid sequence of striped bass pituitary GnRH-R. Based on the hydropathy analysis and sequence comparison by Wisconsin Package Version 9.0 (Genetics Computer Group) with other GnRH-R sequences, putative transmembrane domains (TMD) are deduced and shown as underlined amino acids. The presence of 7 TMDs is characteristic of G-protein coupled receptor family. cDNA sequence was submitted to the gene bank (accession no. AF218841).

The cloned fish GnRH-Rs have invariably shown selectivity for cGnRH II that induces the highest second messenger generation (Tensen et al., 1997; Illing et al., 1999; Alok et al., 2000; Troskie et al., 2000). The GnRH-Rs are G-protein coupled receptors (Fig. 7.5) which, upon activation, can trigger multiple signal transduction pathways (STPs). GnRH-R present on the gonadotrophs mainly transduces its signals via activation of phospholipase C in the gonadotrophs of catfish, goldfish and Rat $\propto T_3$ cells, resulting in the production of Ca^{++}/inositol-phosphotase (IP) for LH synthesis and release (Rebers et al., 2000). In the dispersed pituitary cells of a perciform fish (tilapia), cross-talk between both cAMP-PKA and PKC pathways has been suggested in the GnRH-R-mediated effect on the LH (Melamed et al., 1998). GnRH-R can mediate GnRH functions via cAMP production in sf9 cells (Delahaye et al., 1997) and rat GH3 cells (Lin and Conn, 1999).

These in vitro studies will streamline some of the areas of second messenger generation leading to the release of LH. However, monitoring the second messengers in the dispersed pituitary cells is essential (Stefano et al., 1999; Gur et al., 2001;) the mixture of cells (gonadotropes, somatotropes etc) may be activating their own specific STPs. It appears that in the absence of some signal(s) in captivity, the endogenous GnRHs are activating the STPs, which lead to the arrest of vitellogenesis, FOM and ovulation in most of the studied fish species.

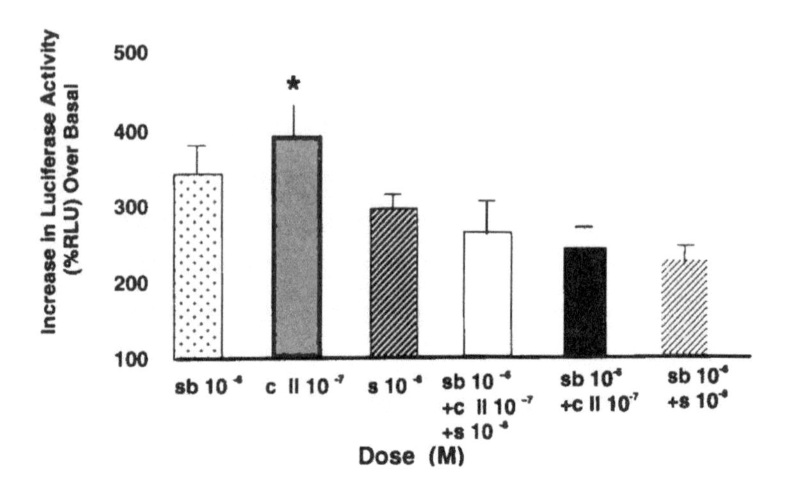

Fig. 6.6: Recombinant striped bass pituitary GnRH-R-mediated *c-fos*-driven luciferase activities (RLU) in CHSE 214 fish cells in response to different doses of the individual native GnRHs [seabream (sb), chicken (c) and salmon (s)] and their combinations. Asterisks (*) on the bar show significant difference ($p < 0.05$) in the luciferase activity. The ratio of the doses of endogenous GnRH combinations were maintained as observed in an *in vivo* situation (Alok et al., 2001).

Therefore, elucidating the mGnRHa-triggered signaling mechanism in the pituitary of farmed fish is necessary to understand mGnRH-based spawning induction technologies. It will enhance our ability to tailor the treatment regimen for future spawning therapies and simplify the selection and utilization of new aquaculture species. Emerging evidence also suggests that endogenous GnRHs may activate distinct signaling pathways (Lovejoy et al., 1995; Johnson et al., 1999). It was demonstrated that the activity of recombinant stbGnRH-R was attenuated in response to the combination of endogenous GnRHs (Fig. 6.6; Alok et al., 2001). The *in vitro* GnRH-R binding and activation studies (with the native GnRHs alone or in combination) in the heterologous systems and dispersed pituitary cells can indicate whether a cocktail of the native forms or their agonists are essential to activate the relevant second messenger generation, leading to the synthesis and release of FSH and LH.

Despite the ability to design and synthesize highly resistant and potent GnRH agonists (Karten and Rivier, 1986; Goren et al., 1990; Zohar et al., 1990b, 1995a; Ngamvonchon et al., 1992b; Alok et al., 1999), screening of GnRH agonists for optimum and compatible receptor binding affinity/specificity/signaling has been seriously hindered by the need to harvest hundreds of pituitaries from valuable broodstock in order to perform large-scale binding affinity/specificity/signaling assays. The use of cloned GnRH-R to develop stable cell lines mimicking the in vivo STPs can most certainly be used to screen GnRH agonists or their combinations. Similar assays have been used to screen GnRH peptides for cancer therapies (Beckers et al. 1997). GnRHas or their selected combinations will be able to activate natural and physiologically compatible signaling in the pituitary of many species of interest and induce predictable spawning.

It is now clear that a fish species contains two or three different types of GnRHs in their brain and they are delivered in differential amounts to the pituitary. The species-specific form is the most dominant form present, followed by cGnRH II and sGnRH. All endogenous GnRHs are LH secretogouge, with cGnRH II being the most potent form. Captivity has adverse effects on the endogenous GnRH system, resulting in the blockage of LH release from the pituitary and arresting the process of vitellogenesis, FOM and spawning in a number of farmed fish species. Now, it is possible to monitor subtle changes in the levels of each of the GnRH forms and GnRH-R forms, due the availability of genes and cDNA sequences in farmed fish (Alok et al., 2000; Wong et al., 2000). Further study is required to determine : (1) whether there are captivity-induced changes in the pituitary GnRH-R number, specificity, affinity, and synthesis during the reproductive cycle in the species of interest; (2) the specific changes in the signaling pathways in the pituitary of captive fish in comparison to the wild fish undergoing FOM; and (3) the specific GnRH treatment needed in captive fish (i.e. which native GnRH forms, alone or in combination), to induce changes similar to those observed in the pituitaries of wild fish undergoing FOM. It appears that endogenous species-specific GnRH (e.g. sbGnRH and cfGnRH) and cGnRH II have been overlooked in the design and development of current spawning induction therapies. The next generation of treatment regimens for induced spawning in farmed fish will be carefully tailored according to the number, specificity, and affinity of the receptor using the most physiologically relevant and endogenous GnRH forms or their combinations. The spawning technology thus developed will, therefore, activate the STPs in the pituitary that most closely mimic the cellular milieu described in the wild fish undergoing FOM. Also, a better insight on the impact of captivity on the fish endocrine reproductive web and on the GnRH and GnRH-R gene regulation offers promise for the development of transgenic fish using specific GnRH forms or GnRH-R that can be turned 'on or off' to initiate gametogenesis, ovulation, spermiation and spawning in a timely manner.

CONCLUSION

Substantial information on the environmental manipulation combined with hormonal treatment has allowed successful interventions in the teleost BPG web to advance, delay or prolong the reproductive cycles of farmed fish. GnRH-based spawning technologies have become a method of choice to induce spawning in a variety of farmed fish. Concomitant with the recent discoveries of multiple GnRHs, the novel GnRH-Rs, and their respective genes in individual species, current spawning technologies can be further improved to meet the challenges of diversification, propagation and conservation in modern aquaculture. Understanding the influence of captivity on the endogenous

GnRH-GnRH-R interactions that block the gonadotropin release and elucidation of the relevant cellular signaling will pave the way for the design of novel spawning technologies tailored to meet the endocrinological needs of the species of choice. However, it is important to emphasize that spawning induction is not only administration of hormones to mimic the cellular signaling in the pituitary, but also a combination of efforts to improve the husbandry and selection of maturing fish, lower densities, optimize water quality and enhance broodstock diet with respect to the individual species requirements.

ACKNOWLEDGEMENTS

The authors thank John Stubblefield for his editorial help. We are also grateful to the National Science Foundation of USA, Maryland and Massachusetts Sea Grant programs, Maryland Agriculture Experiment Station and United States Department of Agriculture for funding separate aspects of research described in this review.

REFERENCES

Alok, D., Krishnan, T., Talwar, G.P. and Garg, L.C. 1993. Induced spawning of catfish, *Heteropneustes fossillis* by D-Lys^6 salmon gonadotropin releasing hormone analog. Aquaculture 115: 159-167.

Alok, D., Pillai, D., Talwar, G.P. and Garg, L.C. 1995. D-Lys^6 salmon gonadotropin-releasing hormone analog-domperidone induced ovulation in *Clarias batrachus*. Asian Fish. Sci. 8: 263-266.

Alok, D., Pillai, D., Talwar, G.P. and Garg, L.C. 1997. Effects of D-Lys^6 salmon gonadotropin releasing hormone analog alone and in combination with domperidone on the spawning of common carp (*Cyprinus carpio*). Aquacult. Internatl. 5: 369-374.

Alok, D., Talwar, G.P. and Garg, L.C. 1999. *In vivo* activity of the native chicken gonadotropin-releasing hormone (cGnRH-II), salmon gonadotropin-releasing hormone and its analogs (sGnRHa) with variable amino acid residues in positions 6, 9, and 10 on the spawning of catfish, *Heteropneustes fossilis* (Bloch). Aquacult. Internatl. 7: 383-392.

Alok, D., Hassin, S., Kumar R.S., Trant, J.M., Yu K.L. and Zohar Y., 2000. Characterization of a pituitary GnRH-receptor from a perciform fish, *Morone saxatilis*: Functional expression in a fish cell line. Mol. Cell. Endocrinol. 168: 65-75.

Alok, D., Kumar R.S., Trant, J.M. and Zohar Y. 2001. Recombinant perciform GnRH-R expressed in a fish and mammalian heterologous cell lines activate differential signaling. Comp. Biochem. Physiol. 129: 375-380.

Beckers T., Reilander, H. and Hilgard, P. 1997. Characterization of gonadotropin releasing hormone analogs based on a sensitive cellular luciferase reporter gene assay. Anal. Biochem. 251: 17-23.

Bauer-Dantoin, A.C., Hollenberg, A.N., Jameson, J.L. 1993. Dynamic regulation of gonadotropin-releasing hormone receptor mRNA levels in the anterior pituitary gland during the rat estrous cycle. Endocrinology 133: 1911-1914.

Blaise, O., Mananos, E. and Zohar, Y. 1996. Development and validation of a radioimmunoassay for studying plasma levels of gonadotropin II (GtH-II) in striped bass (*Morone saxatilis*). Annales d'Endocrinologie. 57: 65.

Blythe, W., Helfrich, L.A., and Sullivan, C.V. 1994. Sex steroid hormone and vitellogenin levels in striped bass (*Morone saxatilis*) maturing under 6-, 9-, and 12-month photothermal cycles. Gen. Comp. Endocrinol. 94: 122-134.

Breton, B., Weil, C., Sambroni, E. and Zohar, Y. 1990. Effects of acute versus sustained administration of GnRHa on GtH release and ovulation in the rainbow trout, *Oncorhynchus mykiss*. Aquaculture. 91: 371-383.

Bromage, N., Jones, J., Randall, C., Thrush, M., Springate, J., Duston, J. and Barker J. 1992. Broodstock management, fecundity, egg quality and the timing of egg production in the rainbow trout, (*Oncorhynchus mykiss*). Aquaculture 100: 141-166.

Brooks, J., Taylor, P.L., Saunders, P.T.K., Eidene, K., Struthers, W.J., McNeilly, A.S. 1993. Cloning and sequencing of the sheep pituitary gonadotropin-releasing hormone receptor and changes in expression of its mRNA during the estrus cycle. Mol. Cell. Endocrinol. 94: R23-R27.

Burgus, R., Butcher, M., Amoss, M., Ling, N., Monahan, M., Rivier, J., Fellows, R., Blackwell, R., Vale, W. and Guillemin, R. 1972. Primary structure of the ovine hypothalamic luteinizing hormone releasing factor (LRF). Proc. Natl. Acad. Sci. USA 69: 278-284.

Carolsfeld, J., Powell, J.F., Park, M., Fischer, W.H., Craig, A.G., Chang, J.P., Rivier, J.E., Sherwood, N.M. 2000. Primary structure and function of three gonadotropin-releasing hormones, including a novel form, from an ancient teleost, herring. Endocrinology, 141 (2), 505–512.

Carolsfeld, J., Sherwood, N.M., Kreiberg, H., Sower, S.A., 1988. Induced sexual maturation of herring using GnRHa 'quick-release' cholesterol pellets. Aquaculture 70, 169-181.

Chang, J.P., Van Goor, F., Jobin, R.M. and Lo, A. 1996. GnRH signaling in goldfish pituitary cells. Biol. Signals. 5: 70-80.

Chasin, M. and Langer, R. 1990. Biodegradable polymers as drug delivery systems. In: Drugs and the Pharmaceutical Sciences, (ed.) Swarbrick, J., Dekker, Marcel New York, 45: 347-365.

Chow, M., Kight, K., Gothilf, Y., Alok, D. and Zohar, Y. 1998. Multiple GnRHs present in a teleost species are encoded by separate genes: Analysis of the sbGnRH and cGnRH II genes from the striped bass, *Morone saxatilis*. J. Mol. Endocrinol. 21: 277-289.

Coe, I.R, von Schalburg, K.R, Sherwood, N.M. 1995. Characterization of the Pacific salmon gonadotropin-releasing hormone gene, copy number and transcription start site. Mol. Cell. Endocrinol. 115: 113-22.

Crim, L.W., St. Arnaud, R., Lavoie, M. and Labrie, F. 1988. A study of LHRH receptors in the pituitary gland of the winter flounder (*Pseudopleuronectes americanus* Walbaum). Gen. Comp. Endocrinol. 69: 372-377.

Delahaye, R., Manna, P.R., Berault, A., Berreur-Bonnenfant, J., Berreur, P. and Counis, R. 1997. Rat gonadotropin-releasing hormone receptor expressed in insect cells induces activation of adenylyl cyclase. Mol. Cell Endocrinol. 135: 119-127.

De Leeuw, R., Goos, H.J.T., Richter, C.J.J., and Eding, E.H. 1985. Pimozide-LHRH-a induced breeding in the African catfish, *Clarias gariepinus* (Burchell). Aquaculture 44: 295-302.

De Leeuw, R, Conn, M.P., van't Veer, C., Goos, H.J.Th. and van Oordt, P.G.W.J. 1988. Characterization of the receptor for gonadotropin-releasing hormone in the pituitary of the African catfish, *Clarias gariepinus*. Fish Physiol. Biochem. 5: 99-107.

Fermin, C.A. 1991. LHRH-a and domperidone-induced oocyte maturation and ovulation in bighead carp, *Aristichthys nobilis* (Richardson). Aquaculture 93: 87-94.

Genazzani, A.R., Bernardi, F., Monteleone, P., Luisi, S., Luisi, M. 2000. Neuropeptides, neurotransmitters, neurosteroids, and the onset of puberty. Ann. N. Y. Acad. Sci. 900: 1-9.

Gluvokov, A.I., Motloch, N.N. and Sedova, M.A. 1991. Effect of a synthetic LHRH analogue and dopamine antagonists on the maturation of bream, *Abramis brama* L. Aquaculture 95, 373–377.

Goren, A., Zohar, Y., Fridkin, E., Elhanati, E. and Koch, Y. 1990. Degradation of gonadotropin-releasing hormone in the gilthead seabream, *Sparus aurata*: Cleavage of native salmon GnRH, mammalian LHRH and their analogs in the pituitary. Gen. Comp. Endocrinol. 79: 291-305.

Gothilf, Y. and Zohar, Y. 1991. Clearance of different forms of GnRH from the circulation of gilthead seabream, *Sparus aurata*, in relation to their degradation and bioactivity. In: *Reproductive Physiology of Fish* (eds) A.P. Scott, J.P. Sumpter, D.E. Kime and M.S. Rolfe, University of East Anglia Press. pp. 35-37.

Gothilf, Y., Elizur, A. and Zohar, Y. 1995a. Three forms of gonadotropin-releasing hormone in gilthead seabream and striped bass: physiological and molecular studies. In: (eds) F.W. Goetz and P. Thomas, Proceedings of the Fifth International Symposium on Reproductive Physiology of Fish, July 2-8. Austin, Texas.

Gothilf, Y., Elizur, A., Chow, M., Chen, T.T. and Zohar, Y. 1995b. Molecular cloning and characterization of a novel gonadotropin-releasing hormone from the gilthead seabream (*Sparus aurata*). Mol. Mar. Biol. Biotech. 4: 27-35.

Gothilf, Y., Munoz-Cueto, J.A., Sagrillo, C.A., Selmanoff, M., Chen, T.T., Kah, O., Elizur, A., and Zohar, Y. 1996. Three forms of gonadotropin-releasing hormone in a perciform fish (*Sparus aurata*): Complementary deoxyribonucleic acid characterization and brain localization. Biol. Reprod. 55: 636-645.

Gothilf, Y., Meiri, I., Elizur, A. and Zohar, Y. 1997. Changes in mRNA levels of the three GnRH forms during final oocyte maturation, ovulation and spawning in the gilthead seabream, *Sparus aurata*. Biol. Reprod. 57: 1145-1154.

Groves, D.J., Batten, T.F. 1986. Direct control of the gonadotroph in a teleost, *Poecilia latipinna*. II. Neurohormones and neurotransmitters. Gen. Comp. Endocrinol. 62. 315-326.

Gur, G., Bonfil, D., Safarian, H., Naor, Z., Yaron, Z. 2001. GnRH receptor signaling in tilapia pituitary cells: role of mitogen-activated protein kinase (MAPK). Comp. Biochem. Physiol. B Biochem. Mol. Biol. 129: 517-24.

Habibi, H.R., Peter, R.E., Sokolowska, M., Rivier, J. and Vale, W.W. 1987. Characterization of gonadotropin-releasing hormone (GnRH) binding to pituitary receptors in goldfish (*Carassius auratus*). Biol. Reprod. 36: 844-853.

Habibi, H.R., Marchant, T.A., Nahorniak, C.S., Van Der Loo, H., Peter, R.E., Rivier, J.E, and Vale, W.W. 1989. Functional relationship between receptor binding and biological activity for analogs of mammalian and salmon gonadotropin-releasing hormone in the pituitary of goldfish (*Carassius auratus*). Biol. Reprod. 40: 1152-1161.

Hassin, S., Monbrison, de., Hanin, Y., Zohar, Y. and Popper, D.M. 1997. Domestication of the white grouper, *Epinephelus aeneus* : I. Growth and reproduction. Aquaculture, 156: 305-316.

Hodson, R.G. and Sullivan, C.V. 1993. Induced maturation and spawning of domestic and wild striped bass, *Morone saxatilis* (Walbaum), broodstock with implanted GnRH analogue and injected hCG. Aquacult. Fish. Manag. 24: 389-398.

Holland, M.C.H., Gothilf, Y., Meiri, I., King, J.A., Okuzawa, K., Elizur, A. and Zohar, Y. 1998. Levels of the native forms of GnRH in the pituitary of the gilthead seabream, *Sparus aurata*, at several characteristic stages of the gonadal cycle. Gen. Comp. Endocrinol. 112: 394-405.

Holland, M.C.H., Hassin, S. and Zohar, Y. 2001. Seasonal fluctuations in pituitary levels of the three native GnRH forms in striped bass, *Morone saxatilis* (Teleostei), during juvenile and pubertal development. J. Endocrinol. 169: 527-538.

Illing, N., Troskie, B.E., Nahorniak, C.S., Hapgood, J.P., Peter, R.E. and Millar, R.P. 1999. Two gonadotropin-releasing hormone receptor subtypes with distinct ligand selectivity and differential distribution in brain and pituitary in the goldfish (*Carassius auratus*). Proc. Natl. Acad. Sci., USA. 96: 2526-2531.

Johnson, J.D., Van Goor, F., Wong, C.J., Goldberg, J.I. and Chang, J.P. 1999. Two endogenous gonadotropin-releasing hormones generate dissimilar Ca^{2+} signals in identified goldfish gonadotropes. Gen. Comp. Endocrinol. 116: 178-191.

Kah, O., Anglade, I., Lepretre, E., Dubourg, P., and de Monbrison, D. 1993. The reproductive brain in fish. Fish. Physiol. Biochem. 11: 85-98.

Kakar, S.S., Rahe, C.H. and Neill, J.D. 1993. Molecular cloning, sequencing and characterizing the bovine receptor for gonadotropin-releasing hormone (GnRH). Biochem. Biophys. Res. Commun. 10: 335-342.

Kakar, S.S., Grantham, K., Musgrove, L.C., Devor, D., Sellers J.C., and Neill, J.D. 1994. Rat gonadotropin-releasing hormone (GnRH) receptor: tissue expression and hormonal regulation of its mRNA. Mol. Cell. Endocrinol. 101, 151-157.

Karten, M.J. and Rivier, J.E. 1986. Gonadotropin-releasing hormone analog design. Structure-function studies toward the development of agonists and antagonists: rationale and perspective. Endoc. Rev. 7: 44-66.

Kaul M. and Rishi, K.K., 1986. Induced spawning of the Indian major carp, *Cirrhina mrigala* (Ham.), with LHRH analogue or pimozide. Aquaculture. 54: 45-48.

King, J.A. and Millar, R.P. 1995. Evolutionary aspects of gonadotropin-releasing hormone and its receptor. Cell. Mol. Neurobiol. 15: 5-23.

Lescheid, D.W., Terasawa, E, Abler, L.A., Urbanski, H.F., Warby, C.M., Millar, R.P., Sherwood, N.M. 1997. A second form of gonadotropin-releasing hormone (GnRH) with characteristics of chicken GnRH-II is present in the primate brain. Endocrinology 138: 5618-5629.

Lin, H.R., Peng, C., Van Der Kraak, G., Peter, R.E., and Breton, B. 1986. Effects of [D-Ala^6, Pro^9-NEt]-LHRH and catecholaminergic drugs on gonadotropin secretion and ovulation in the Chinese loach (*Paramisgurnus dabyranus*). Gen. Comp. Endocrinol. 64: 389-395.

Lin, H.R., Kraak, V. der, Zhou, X.J., Liang, J.Y., Peter, R.E., Rivier, J.E and Vale, W.W. 1988. Effects of [D-Arg6, TrP7, Leu8 Pro9]-luteinizing hormone releasing hormone (sGnRH-A) and [D-Ala6, Pro9]-luteinizing hormone releasing hormone (mLHRH-A) in combination with pimozide or domperidone, on gonadotropin release and ovulation in the Chinese loach and common carp. Gen. and Comp. Endocrinol. 69: 31–40.

Lin, X., Janovick, J.A., Brothers, S., Blomenrohr, M., Bogerd, J. and Conn, P.M. 1998. Addition of catfish gonadotropin-releasing hormone (GnRH) receptor intracellular carboxyl-terminal tail to rat GnRH receptor alters receptor expression and regulation. Mol. Endocrinol. 12: 161-171.

Lin, X. and Conn, P.M. 1999. Transcriptional activation of gonadotropin-releasing hormone (GnRH) receptor gene by GnRH: Involvement of multiple signal transduction pathways. Endocrinology 140: 358-364.

Lovejoy, D.A., Corrigan, A.Z., Nahorniak, C.S., Perrin, M.H., Porter, J., Kaiser, R., Millar., C., Pantoja, D., Craig, A.G., Peter, R.E., Vale, W. W., Rivier, J.E., and Sherwood, N.M. 1995. Structural modifications of non-mammalian gonadotropin-releasing hormone (GnRH) isoforms: Design of novel GnRH analogues. Reg. Pept. 60: 99-115.

Madigou, T., Manaños-Sanchez, E., Hulshof, S., Anglade, I., Zanuy, S. and Koh, O. 2000. Cloning, tissue distribution, and central expression of the gonadotropin-releasing hormone receptor in the Rainbow trout (*Oncorhynchus mykiss*). Biol. Reprod. 63:1857-1866.

Mason, D.R., Arora, K.K., Mertz, L.M. and Catt, K.J. 1994. Homologous downregulation of gonadotropin-releasing hormone receptor sites and messenger ribonucleic acid transcripts in alpha T3-1 cells. Endocrinology 135: 1165-1170.

Matsuo, H., Baba, Y., Nair, R.M.G., Arimura, A. and Schally, A.V. 1971. Structure of the porcine LH and FSH releasing hormone. I. The proposed amino acid sequence. Biochem. Biophys. Res. Commun. 43: 1334-1339.

Melamed, P., Rosenfeld, H., Elizur, A. and Yaron, Z. 1998. Endocrine regulation of gonadotropin and growth hormone gene transcription in fish. Comp. Biochem. Physiol. C Pharmacol. Toxicol. Endocrinol. 119: 325-338.

Millar R., Lowe, S. Conklin, D., Pawson, A., Maudsley, S., Troskie, B., Ott, T., Millar, M., Lincoln, G., Sellar, R., Faurholm, B., Scobie, G., Kuestner, R., Terasawa, E. and Katz, A. 2001. A novel mammalian receptor for the evolutionarily conserved type II GnRH. Proc. Natl. Acad. Sci., USA. 98: 9636-9641.

Montero, M., Vidal, B., King, J.A., Tramu, G., Vandesande, F., Dufour, S. and Kah, O. 1994. Immunocytochemical localization of mammalian GnRH (gonadotropin- releasing hormone) and chicken GnRH-II in the brain of the European silver eel (*Anguilla anguilla* L.). J. Chem. Neuroanat. 7: 227-241.

Mylonas, C.C., Tabata, Y., Langer, R. and Zohar, Y. 1995a. Preparation and evaluation of polyanhydride microspheres containing gonadotropin-releasing hormone (GnRH), for inducing ovulation and spawning in fish. J. Cont. Rel. 35: 23-34.

Mylonas, C.C., Zohar, Y., Richardson, B.M. and Minkinnen, S.P. 1995b. Induced spawning of American shad (*Alosa sapidissima*) using sustained administration of gonadotropin-releasing hormone analog (GnRHa). J. World Aquacult. Soc. 26: 39-50.

Mylonas, C.C., Magnus, Y., Gissis, A., Klebanov, Y. and Zohar, Y. 1996. Application of controlled-release, GnRHa delivery systems in commercial-scale production of white bass x striped bass hybrids (sunshine bass), using captive broodstocks. Aquaculture 140: 265-280.

Mylonas, C.C., Woods, L.C. and Zohar, Y. 1997a. Cyto-histological examination of post-vitellogenesis and final oocyte maturation in captive-reared striped bass (*Morone saxatilis* Walbaum). J. Fish Biol. 50: 34-49.

Mylonas, C.C., Gissis, A., Magnus, Y. and Zohar, Y. 1997b. Hormonal changes in male white bass (*Morone chrysops*) and evaluation of milt quality after treatment with sustained release GnRHa-delivery system. Aquaculture 153: 301-313.

Mylonas, C.C., Magnus, Y., Gissis, A. and Zohar, Y. 1997c. Reproductive biology and endocrine regulation of final oocyte maturation of captive-reared white bass (*Morone chrysops* Rafinesque). J. Fish Biol. 51: 234-250.

Mylonas, C.C., Scott, A.P., Vermeirssen, E.L.M. and Zohar, Y. 1997d. Changes in plasma gonadotropin II and sex-steroid hormones, and sperm production of striped bass after treatment with controlled-release GnRHa-delivery systems. Biol. Reprod. 57: 669-675.

Mylonas, C.C., Scott, A.P. and Zohar, Y. 1997e. Plasma gonadotropin II, sex steroids and thyroid hormones in wild striped bass (*Morone saxatilis*) during spermiation and final oocyte maturation. Gen. Comp. Endocrinol. 108: 223-236.

Mylonas. C.C. and Zohar, Y. 1998. New technologies for the control of gamete maturation in marine fishes, as tools in broodstock management. In: Cahiers Options Mediterraneennes, Genetics and Breeding in Mediterranean Aquaculture Species, CIHEAM, Spain, 34: 123-157.

Mylonas, C.C., Woods, L.C., Thomas, P. and Zohar, Y. 1998. Endocrine profiles of female striped bass (*Morone saxatilis*) during post-vitellogenesis and induction of final oocyte maturation via controlled-release GnRHa-delivery systems. Gen. Comp. Endocrinol. 110: 276-289.

Nandeesha, M.C., Keshavnath, P., Verghese, T.J., Shetty, H.P.C., and Rao, K.G. 1990. Alternate inducing agents for carp breeding: progress in research. In: Carp seed production technology (eds) P. Keshavanath and K.V. Radhakrishnan. Proc. workshop carp seed Prod.Technol. Asian fisheries society, Indian branch, Mangalore, India pp. 12-16.

Neil, J.D., Duck, L.W., Sellers, J.C. and Musgrove, L.C. 2001. A gonadotropin-releasing hormone receptor specific for GnRH II for primates. Biochem. Biophys. Res. Commun. 282: 1012-1018.

Ngamvonchon, S., Lovejoy, D.A., Fischer, W.H., Craig, A.G., Nahorniak, C.S., Peter, R.E., Rivier, J.E., and Sherwood, N.M. 1992a. Primary structures of two forms of gonadotropin-releasing hormone, one distinct and one conserved form in catfish brain. Mol.Cell. Neurosci. 3: 17-22.

Ngamvonchon, S., Rivier, J.E., and Sherwood, N.M. 1992b. Structure-function studies of five natural, including catfish and dogfish, gonadotropin-releasing hormones and eight analogs on reproduction in Thai catfish, *Clarias macorcephalus*. Reg. Pept. 42: 63-73.

Okubo, K., Masafumi, A., Yoshiura, Y., Suetake, H. and Aida. K. 2000a. A novel form of gonadotropin-releasing hormone in the Medaka, *Oryzias latipes*. Biochem. Biophys. Res. Comm. 276: 298-303.

Okubo, K., Suetake, H., Usami, T. and Aida, K. 2000b. Molecular cloning and tissue-specific expression of a gonadotropin-releasing hormone receptor in the Japanese eel. Gen. Comp. Endocrinol. 119: 181-192.

Omeljaniuk, R.J., Habibi, H.R. and Peter, R.E. 1989. Alterations in pituitary GnRH and dopamine receptors associated with the seasonal variation and regulation of gonadotropin release in the goldfish (*Carassius auratus*). Gen. Comp. Endocrinol. 74: 392-399.

Okuzawa, K., Granneman, J., Bogerd, J., Goos, H.J.Th., Zohar, Y. and Kagawa, H. 1997. Distinct expression of GnRH genes in the red seabream brain. Fish Physiol. Biochem. 17: 71-79.

Padmanabhan, V., Dalkin, A., Yasin, M., Haisenleder, D.J., Marshal, J.C. and Landefield, T.D. 1995. Are early genes involved in gonadotropin-releasing hormone receptor gene regulation ? Characterization of changes in GnRH-receptor (GnRH-R), c-fos, and c-jun messenger ribonucleic acids during the ovine estrous cycle. Biol. Reprod. 53: 263-269.

Pagelson, G. and Zohar, Y. 1992. Characterization of gonadotropin-releasing hormone (GnRH) receptors in the gilthead seabream (*Sparus aurata*). Biol. Reprod. 47: 1004-1008.

Parhar, I.S. and Sakuma, Y. 1997. Regulation of forebrain and midbrain GnRH neurons in juvenile teleosts. Fish Physiol. and Biochem. 17: 81-84.

Peter, R.E., Sokolowska, M., Nahorniak, C.S., Rivier J.E. and Vale, W.W. 1987. Comparison of [D-Arg^{6}Trp^{7}Leu^{8}Pro^9] NEt sGnRH-A), and [DAla^6, Pro^9 NEt.] LHRH-A, in combination with pimozide in stimulating gonadotropin release and ovulation in the goldfish, *Carassius auratus*. Can. J. Zool. 65: 987-991.

Peter, R.E, Habibi, H.R., Chang, J.P., Nahorniak, C.S., Yu, K.L., Huang, Y.P., Marchant T.A. 1990. Actions of gonadotropin-releasing hormone (GnRH) in the goldfish. Prog. Clin. Biol. Res. 342: 393-398

Pawson, A.J., Katz, A., Sun, Y.M., Lopes, J., Illing, N., Millar, R.P. and Davidson, J.S. 1998. Contrasting internalization kinetics of human and chicken gonadotropin-releasing hormone receptors mediated by C-terminal tail. J. Endocrinol. 156: R9-12.

Powell, J.F.F., Zohar, Y., Elizur, A., Park, C., Fischer, W.H., Craig, A.G., Rivier, J.E., Lovejoy, D.A., and Sherwood, N.M. 1994. Three forms of gonadotropin-releasing hormone characterized from brain of one species. Proc. Natl. Acad. Sci., USA. 91: 12081-12085.

Powell, J.F.F., Fischer, W.H., Park, M., Craig, A.G., Rivier, J.E., White, S.A., Francis, R.C.F., Fernald, R.D., Licht, P., Warby, C., and Sherwood, N.M. 1995. Primary structure of a solitary form of gonadotropin-releasing hormone (GnRH) in cichlid pituitary: Three forms of GnRH in brain of cichlid and pumpkinseed fish. Reg. Pept. 57: 43-53.

Powell, J.F.F., Standen, E.M., Carolsfeld, J., Borella, M.I., Gazola, R., Fischer, W.H., Park, M., Craig, A.G., Warby, C.M., Rivier, J.E., Val-Sella, M.V. and Sherwood, N.M. 1997. Primary structure of three forms of gonadotropin-releasing hormone (GnRH) from the pacu brain. Reg. Pept. 68: 189-195.

Ramos, J. 1986. Luteinizing hormone-releasing hormone analogue (LH-RHa) induced precocious ovulation in common sole (*Solea solea* L.). Aquaculture. 54: 185-190.

Rebers, F.E.M., Bosma, P.T., Dijk, W.V., Goos, H.J.T. and Schulz, R.W. 2000. GnRH stimulates LH release directly via inositol phosphate and indirectly via cAMP in African catfish. Am. J. Physiol. 278: R1572-R1578.

Robison, R.R., White, R.B., Illing, N., Troskie, B.E., Morley, M., Millar, R.P., and Fernald, R.D. 2001. Gonadotropin-releasing hormone receptor in the teleost *Haplochromis burtoni*: Structure, location, and function. Endocrinology 142: 1737-1743.

Sealfon, S.C., Weinstein, H. and Millar, R.P. 1997. Molecular mechanisms of ligand interaction with the gonadotropin-releasing hormone receptor. Endocrinol. Rev. 18: 180-205.

Sherwood, N., Eiden, L., Brownstein, M., Spiess, J., Rivier, J. and Vale, W. 1983. Characterization of a teleost gonadotropin-releasing hormone. Proc. Natl. Acad. Sci. USA. 1983 80(9): 2794-2798.

Sherwood, N.M., Parker, D.B., McRory, J.E. and Lescheid D.W. 1994. Molecular evolution of growth hormone-releasing hormone and gonadotropin-releasing hormone. In: Molecular Endocrinology of Fish—Fish Physiology volume, XIII. pp. 3-67. (eds) Farrell, A.P. and Randall, D.J. Academic Press, San Diego.

Sherwood, N.M., vonSchalburg, K. and Lescheid, D.W. 1997. Origin and evolution of GnRH in vertebrates and invertebrates. In: *GnRH Neurons: Gene to Behavior.* (eds) Parhar S. and Sakuma, Y. Brain Shuppan Publishing; Tokyo; pp. 3-25.

Somoza, G.M. and Peter, R.E. 1991. Effects of serotonin on gonadotropin and growth hormone release from in vitro perifused goldfish pituitary fragments. Gen. Comp. Endocrinol. 82: 103-10.

Stefano, V.A., Vissio, P.G., Paz, D.A., Somoza, G.M., Maggese, M.C. and Barrantes, G.E. 1999. Colocalization of GnRH binding sites with gonadotropin-, somatotropin-, somatolectin-, and prolactin-expressing pituitary cells of the pejerry, *Odontethes bonariensis*, in vitro. Gen. Comp. Endocrinol. 116: 133-139.

Steven, C.R. 1999. Studies of the GnRH/GtH system of female striped bass: Effects of GnRH agonist therapy and comparison of reproductive endocrine parameters between wild and captive fish undergoing final oocyte maturation. M.S. dissertation, University of Maryland, USA.

Steven, C.R. , Gothilf, Y., Alok, D. and Zohar, Y. 2000. Comparison of gnrh/gth/gonadal profiles between wild and captive striped bass (*Morone saxatilis*) undergoing final stages of oogenesis. Proceedings of the 4th international symposium on Fish Endocrinology in Seattle.

Stojilkovic, S.S., Reinhart, J. and Catt, K.J. 1994. Gonadotropin-releasing hormone receptors: Structure and signal transduction pathways. Endocrine Rev., 15: 462-499.

Tensen, C., Okuzawa, K., Blomenroehr, M., Rebers, F., Leurs, R., Bogerd, J., Schulz, R. and Goos, H. 1997. Distinct efficacies for two endogenous ligands on a single cognate gonadoliberin receptor. Eur. J. Biochem. 243: 134-140.

Tharakan, B. and Joy, K.P. 1996. Effects of mammalian gonadotropin-releasing hormone analogue, pimozide and the combination on plasma gonadotropin levels in different seasins and induction of ovulation in female catfish. Aquaculture. 48: 623-632.

Troskie, B.E., Hapgood, J.P., Millar, R.P. and Illing, N. 2000. Complementary deoxyribonucleic acid cloning, gene expression, and ligand selectivity of a novel gonadotropin-releasing hormone receptor expressed in the pituitary and midbrain of *Xenopus laevis*. Endocrinology, 141: 1764-1771.

Turzillo, A.M., Campion C.E., Clay C.M. and Nett, T.M. 1993. Regulation of gonadotropin-releasing hormone receptor (GnRH-R) messenger ribonucleic acid and GnRH receptors during the early pre-ovulatory period in the ewe. Endocrinology 135: 1911-1914.

Wang, L., Boger, J., Choi, H.S., Seong, J.Y., Soh, J.M., Chun, S.Y., Blomenrohr, M., Troskie, B.E., Millar, R.M., Yu, W.H., McCann, S.M. and Kwon, H.B. 2001. Three distinct types of GnRH receptor characterized in the bullfrog. Proc. Natl. Acad. Sci. USA. 98: 361-366.

Weber, G.M., Powell, J.F., Park, M., Fischer, W.H., Craig, A.G., Rivier, J.E., Nanakorn, U., Parhar, I.S., Ngamvongchon, S., Grau, E.G. and Sherwood, N.M. 1997. Evidence that gonadotropin-releasing hormone (GnRH) functions as a prolactin-releasing factor in a teleost fish (*Oreochromis mossambicus*) and primary structures for three native GnRH molecules. J. Endocrinol. 155: 121-132.

White, R.B. and Fernald R.D. 1998. Genomic structure and expression sites of three gonadotropin-releasing hormone genes in one species. Gen. Comp. Endocrinol. 112: 17-25.

Willars, G.B., Heding, A., Vrecl, M., Sellar, R., Blomenrohr, M., Nahorski, S.R. and Eidne, K.A., 1999. Lack of a C-terminal tail in the mammalian gonadotropin-releasing hormone receptor confers resistance to agonist-dependent phosphorylation and rapid desensitization. J. Biol. Chem. 274: 30146-30153.

Woods, L.C., III and Sullivan, C.V. 1993. Reproduction of striped bass, *Morone saxatilis* (Walbaum), broodstock: Monitoring maturation and hormonal induction of spawning. Aquacult. Fish. Manag. 24: 211-222.

Wong, T.-S., Gothilf, Y., Meiri, I., Zmora, N., Elizur, A., and Zohar, Y. 2000. Gene expression of three forms of GnRH at different stages of the gilthead seabream life cycle. 4th Internatl. Symp. Fish Endocrinol. Seattle, Washington. pp. 75.

Yahalom D, Chen A, Ben-Aroya N, Rahimipour S, Kaganovsky E, Okon E, Fridkin M, Koch, Y. 1999. The gonadotropin-releasing hormone family of neuropeptides in the brain of human, bovine and rat: identification of a third isoform. FEBS Lett. 463: 289-294.

Yamamoto, N., Parhar, I.S., Sawai, N., Oka, Y. and Ito, H. 1998. Preoptic gonadotropin-releasing hormone (GnRH) neurons innervate the pituitary in teleosts. Neurosci. Res. 31: 31-38.

Zandbergen, M.A., Kah, O., Bogerd, J., Peute, J. and Goos, H.J. 1995. Expression and distribution of two gonadotropin-releasing hormones in the catfish brain. Neuroendocrinology. 62: 571-578.

Zohar, Y. 1988. Gonadotropin-releasing hormone in spawning induction in teleosts: Basic and applied considerations. In: Reproduction in Fish: Basic and Applied Aspects in Endocrinology and Genetics (eds) Zohar Y. and Breton. B. INRA Press, Paris. pp. 47-62

Zohar, Y. 1989. Fish Reproduction: Its physiology and artificial manipulation. In: Fish culture in warm water systems: Problems and Trends. (eds) Shilo, M. and Sarig, S. CRC Press. Florida. pp. 65-119.

Zohar, Y., Goren, A. Tosky M., Pagelson G., Leibovitz D. and Koch Y. 1989a. The bioactivity of gonadotropin-releasing hormones and its regulation in the gilthead seabream, Sparus aurata: *in vivo* and *in vitro* studies. Fish Physiol. Biochem. 7: 59-67.

Zohar, Y., Tosky, M., Pagelson, G., and Finkelman, Y. 1989b. Induction of spawning in the gilthead seabream, *Sparus aurata*, using [D-Ala6-Pro9NET]-LHRH: Comparison with the use of HCG. Israeli J. Aquaculture 41: 105-113.

Zohar, Y., Pagelson, G., Gothilf, Y., Dickhoff, W.W., Swanson, P., Duguay, S., Gombotz, W., Kost J., and Langer, R. 1990a. Controlled release of gonadotropin-releasing hormones in farmed fish. Proc. Inter. Symp. Control. Rel. Bioact. Mater. 17: 51-52.

Zohar, Y., Goren, A., Fridkin, M., Elhanati, E. and Koch, Y. 1990b. Degradation of gonadotropin releasing hormone in the gilthead seabream, *Sparus aurata*: II. Cleavage of native salmon GnRH, mammalian LHRH and their analogs in the pituitary, kidney, and liver. Gen. Comp. Endocrinol. 79: 306-319.

Zohar, Y., Elizur, A., Sherwood N.M., Rivier J.F. and Zmora, N. 1995a. Gonadotropin-releasing potencies of the three native forms of gonadotropin-releasing hormones present in the brain of gilthead seabream, *Sparus aurata*. Gen. Comp. Endocrinol. 97: 289-299.

Zohar, Y., Harel, M., Hassin, S. and Tandler, A. 1995b. Broodstock management and manipulation of spawning in the gilthead seabream, *Sparus aurata*. In: Broodstock Management and Egg and Larval Quality (eds.) Bromage, N. and Roberts, R.J. Blackwell Science Press, London. pp. 94-117.

Zohar, Y. 1996. New approaches for the manipulation of ovulation and spawning in farmed fish. Bull. Natl. Res. Inst. Aquacult. Suppl., 2: 43-47.

Zohar, Y. and Mylonas, C.C. 2001. Endocrine manipulations of spawning in farmed fish: From hormones to genes. Aquaculture 197: 99-136.

Chapter 7

Biotechnology to Improve Reproductive Performance and Larval Rearing in Prawns and Rock Lobsters

K. Wilson, M. Hall, M. Davey, M. Kenway and **D. Coren**
Australian Institute of Marine Science
PMB 3, Townsville, Qld 4810, Australia
Email: k.wilson@aims.gov.au

ABSTRACT

Prawns and rock lobsters are both high value seafood items which make a significant contribution to the total Australian fisheries production. Prawn aquaculture is a well established industry but has yet to develop as a true farming sector for the black tiger prawn, *Penaeus monodon*, as the industry is reliant on wild-caught broodstock. One of the problems besetting the use of wild-caught or domesticated broodstock is induction of spawning, and in this chapter we discuss the hormonal regulation of ovarian maturation and how biotechnological approaches may overcome this blockage.

There is at present no rock lobster aquaculture industry in Australia. The biggest hurdle to overcome in achieving successful commercial aquaculture production of this species group is larval rearing. Good nutrition and health are clearly critical for successful larval rearing. In addition, by understanding the hormonal control of larval development, it may be possible to apply chemical cues to speed up development and shorten the length of the larval phase.

Keywords: Sinus gland peptides, prawns, rock lobsters, spawning, larval, development

INTRODUCTION

In Australia, rock lobsters and prawns dominate wild fisheries production. In 2001-02, the catches were worth $545 m, and $415 m, respectively, a combined value of almost 40% of Australia's total fisheries production (ABARE, 2002). Of these landings, approximately 50% of the prawn harvest and almost all the rock lobster harvest are destined for export, making these fisheries significant contributors to the Australian economy (ABARE, 2002). However, both fisheries are now considered to be fully exploited. Thus, there is no capacity within the wild fisheries of Australia to capitalize on the growing global demand for these products (e.g. Fig. 7.1), which illustrates the demand for imported prawns in the USA.

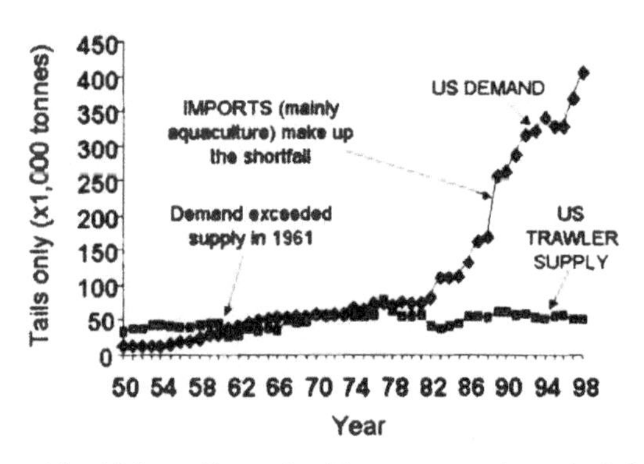

Fig. 7.1: Demand for high quality seafood is consumer dependent. Imports of wild-caught and aquacultured prawns have outstripped the USA wild population harvest to meet consumer demand.

The only means by which Australia can increase production of these valuable crustacean species to meet increasing global demand and earn valuable export income is by aquaculture. Aquaculture of prawns is by now well established throughout Asia and the Americas, and currently provides just over one quarter of total global shrimp landings (FAO 2003) . However, it still remains a young industry beset by problems that cause large regional fluctuations in production. Aquaculture of rock lobsters has not yet been demonstrated to be a real possibility primarily due to the lengthy and complex larval cycle. In this chapter, we present an overview of the work, which is being carried out at the Australian Institute of Marine Science to overcome key bottlenecks in aquaculture production of both prawns and rock lobsters.

PRAWN AQUACULTURE RESEARCH

Challenges facing global prawn aquaculture

The penaeid aquaculture industry in Australia is dwarfed by that in Southeast Asia and China (1.3 million t produced in 2003), of which more than half is

Penaeus monodon (Rosenberry, 2003). By contrast, the Australian industry had a total production of 3,696 t in 2001-2002 (ABARE, 2002). *P. monodon* is the dominant species in Australia, accounting for > 60% of the total value of production. The other species currently farmed in Australia are *P. merguiensis* and *P. japonicus* (O'Sullivan and Dobson, 2000).

Worldwide growth in the penaeid farming industry has been hampered by adverse environmental impacts, disease and shortages of broodstock (see Table 7.1). Many of these problems could be overcome by development of fully closed farming systems. Use of a fully recirculating water system mitigates disease problems and substantially reduces any effluent impacts. These systems are best suited to use with domesticated stock to avoid introduction of wild broodstock of unknown health status.

Table 7.1: Trends in global penaeid prawn production from aquaculture. Values given in brackets represent the changes.

Prawn production metric tons	*1996*	*1997*	*1998*	*1999*
Eastern hemisphere	519,100	462,000 (-11%)	531,200 (+15%)	642,750 (+21%)
Western hemisphere	172,348	198,200 (+15%)	206,626 (+4%)	171,500 (-17%)
Total	691,448	660,200 (-5%)	737,826 (+11.8%)	814,250 (+10.4%)

The penaeid species that have been successfully domesticated on a commercial scale are generally those which mature readily in captivity, often in grow out ponds, including *P. vannamei*, *P. stylirostris* and *P. japonicus* (Preston et al., 1999; Argue et al., 2002; Goyard et al., 2002). Closing the life cycle of *P. vannamei* has facilitated the development of fully closed, biosecure prawn farming systems for this species in Central and South America (McIntosh, 2000).

In contrast, *P. monodon* has proven difficult to domesticate. The animals are harvested from grow-out ponds at 5-6 months as sub-adults and generally do not mature sexually until close to 12 months of age. Keeping large numbers of broodstock in good health for 12 months or longer is a significant challenge for *P. monodon* domestication. There has been some success in closing the life cycle on a small scale, primarily in research institutions, including AIMS and CSIRO in Australia (Kenway and Benzie, 1999; Crocos et al., 1999). However, on a commercial scale, almost all farmed *P. monodon* continue to be produced from wild broodstock, with probably less than 1% from captive broodstock.

Other key factors, which impinge on successful domestication of *P. monodon* include difficulty in achieving reliable spawning of female broodstock, the large variability in spawning performance of captive-reared broodstock, the quality of the spermatophores and the quality of resultant larvae. These factors are currently being investigated in Australia in a major multi-agency collaboration, which involves AIMS, the CSIRO, the Queensland Department of Primary

Industries, three commercial prawn farms, and is driven by the industry body, the Australian Prawn Farmers' Association.

An alternative to eyestalk ablation

Spontaneous spawning of intact broodstock is rare. Preferentially, wild broodstock are captured with nearly completely developed ovaries (i.e. stage III-IV) and these may spawn on the first, second or third nights after arrival in the hatchery. If they do not spawn during this time, artificial stimulation of spawning is required. In addition, artificial spawning may also be required so that animals spawn at the correct time to synchronize larval production with demand from farms.

At present, the only practical means to stimulate spawning is by the removal of one eyestalk (unilateral eyestalk ablation), following which the undeveloped ovary matures over 3-7 days, leading to ovulation and spawning. With wild broodstock, typically 60-80% of prawns spawn. By contrast, the proportion of domesticated broodstock that spawn can be only 20% (Menasveta et al., 1994) or even lower. Although eyestalk ablation may be a reasonably efficient means to induce spawning in wild broodstock, it also results in a number of other physiological effects, due to the physical stress and the removal of other key hormones, which influence the longevity and quality of the broodstock. It is standard practise in *P. monodon* hatcheries to only retain the first and second batches of eggs post ablation, and to reject subsequent spawnings due to a decline in the condition of broodstock and egg numbers and quality. Clearly, there is need for an improved method of inducing spawning in both wild and domesticated *P. monodon*, and other farmed penaeid prawns.

The physiological basis for the effect of eyestalk ablation is removal of the source of a hormone that inhibits ovarian development. This hormone is variously referred to as gonad-inhibiting hormone (GIH) or vitellogenesis-inhibiting hormone (VIH), although the mechanism of action of this hormone is not yet understood.

GIH belongs to a family of closely related hormones that regulate different aspects of prawn physiology. Other hormones in this family are crustacean-hyperglycemic hormone (CHH) which regulates levels of glucose in the haemolymph, and moult-inhibiting hormone (MIH), which regulates the moult cycle. These hormones are all short (70-80 amino acid) peptides with a highly conserved sequence. In particular, they feature six conserved cysteines, which form three intramolecular disulfide bonds (Keller, 1992; Soyez, 1997; Chung et al., 1998).

The CHH family hormones are produced by the paired neurohaemal organs known as the X-organ-sinus gland (XO-SG) complex located in the eyestalk. The X-organ comprises the majority of the cell bodies where the neuropeptides are synthesized and the sinus gland functions as a storage organ from where the hormones are released into the haemolymph (Fingerman, 1992). Although all identified CHH family hormones in Crustacea appear to be produced in the XO-SG complex, immunocytochemistry and in situ hybridization have shown that specialized cells within this complex may produce different members of this hormone family (Dircksen et al., 1988; Laverdure et al., 1994). Moreover, there are also reports of CHH family hormones being synthesized in the central nervous system, and even in the gut, as well as in the eyestalk (De Kleijn et al., 1995; Chang et al., 1999; Chung et al., 1999, Dircksen et al., 2001).

The main strategy that we have been pursuing for achieving replacement of unilateral eyestalk ablation is "passive immunization", i.e. immunization of broodstock animals with an antibody directed against GIH. Our hypothesis is that the use of an antibody that specifically removes GIH circulating in the haemolymph but leaves other hormones unaffected, will induce ovarian development without the associated detrimental effects of eyestalk ablation.

Identification of GIH in *P. monodon*

The strategy of passive immunization requires identification of the *P. monodon* GIH and production of a specific antibody against GIH. The challenge, however, is to determine, which sinus gland peptide is (or are) the GIH. All CHH family hormones are highly homologous to each other and increasing evidence suggests that these neuropeptides are multifunctional. For example, a single hormone in the spider crab *Libinia emarginata* can regulate both synthesis of methyl farnesoate (MF)—which is thought to be a positive effector of gonadal development—and glucose metabolism (Liu and Laufer, 1996). Purified hormones with demonstrated CHH activity also inhibited protein synthesis in ovarian fragments in *P. japonicus* (Khayat et al. 1998).

Nevertheless, proteins do fall into two structural classes (Ohira et al., 1997; Lacombe et al., 1999). Type I precursors consist of a signal sequence and CHH precursor related peptide (CPRP), both of which are sequentially cleaved to release the mature hormone, which is 72-73 amino acids long and is amidated at the C-terminus. Type II precursors also have a signal sequence but lack a CPRP. The mature hormones are generally slightly longer (75-78 residues) and are not amidated at the C-terminus (Fig. 7.2).

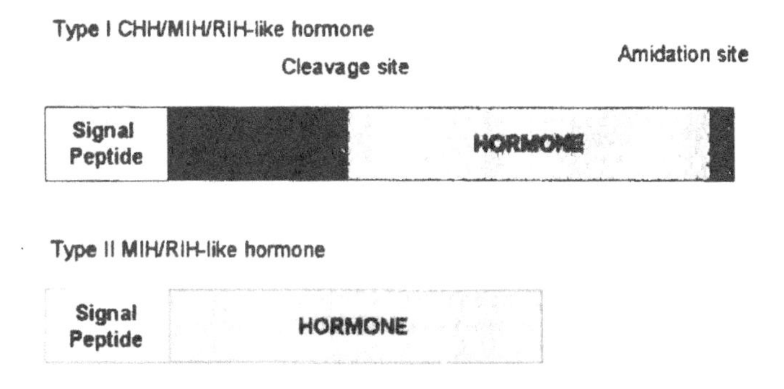

Fig. 7.2: Structure of CHH/MIH/RIH-like hormones

To date, all hormones deemed to be CHHs fall into the Type I class, while hormones specified as MIH, VIH/GIH and MOIH form the known Type II hormones. MOIH stands for Mandibular Organ Inhibiting Hormone, referring to the observation that these hormones, which also originate from the sinus gland, inhibit MF release from the mandibular organ (Wainwright et al. 1996). As it is possible that MF may exert a stimulatory effect on ovarian development, the corollary is that MOIH may be an GIH. MIH and MOIH hormones have also been found with a Type I structure, e.g. the spider crab MOIH (Liu and Laufer, 1996).

One route to obtaining antibodies against putative GIH from *P. monodon* would be to purify the protein(s) from sinus glands and use purified proteins as antigen in antibody production. However, it has not been possible to purify adequate quantities of CHH family hormones due to the extremely small size of the sinus glands in penaeid prawns. For example, purification of proteins in the correct molecular weight class (upto 14 kDa) resulted in only 400 μg of total protein from 491 sinus glands. As about 500 μg of antigen are required per mouse for monoclonal antibody production, and this protein preparation was a mixture of all peptides less than 14 kDa in size from sinus glands, the strategy of producing monoclonal antibodies using purified sinus gland neuropeptides was not practicable.

An alternative strategy was to isolate cDNAs encoding the different sinus gland peptides, express the GIH protein from the corresponding gene and use the expressed protein as antigen for monoclonal antibody production.

Table 7.2: Structure of cDNAs encoding Type I CHH-like hormones from *P. monodon*

PmSGP	*Total cDNA*	*5' UTR*	*Signal peptide*	*CPRP*	*Mature peptide*	*3' UTR*	*Genbank Accession No*
I	813	53	81	57	225	397	AF104386
II	583	48	66	66	225	178	AF104387
III	566	79	66	18	225	178	AF104388
IV	655	33	78	60	225	159	AF104389
V	647	10	78	66	225	268	AF104340

All measurements are in nucleotide lengths. UTR = untranslated region. CPRP = CHH Precursor Related Peptide. The total length of the cDNA is from the presumed transcription initiation to the final residue prior to the poly-A tail. The length given for the coding region for the mature peptide includes the C-terminal two amino acids (GK in all cases) which are cleaved off during the amidation reaction and the termination codon. The 3' UTRs of PmSGPII and PmSGPIII show substantial homology, but there is otherwise little obvious homology between the 5' or 3' UTRs of the respective cDNAs.

To achieve this goal, proteins purified from sinus glands were subjected to Edman degradation to obtain partial amino acid sequence, and were used these sequences to design PCR primers for isolating gene fragments. This work is described by Davey et al., (2000) and has led to the isolation of five different full-length cDNAs (Table 7.2). Conceptual translation of these cDNAs indicated that they all encoded Type I proteins, and predicted exact molecular weights for the mature proteins based on the assumption of both signal peptide and prepropeptide cleavage, C-terminal amidation and formation of three intramolecular disulfide bonds. These predictions were verified using Electrospray Ionization with a Fourier Transform Mass Spectrometer (FTMS), which also indicated clearly that all five hormones, designated PmSGPI - V, were present in single sinus glands of more than 30 animals examined (Davey et al. 2000). Thus, the cDNAs did not represent polymorphisms but indeed 5 different neuropeptides. This is consistent with data from *P. japonicus*, in which six different Type I neuropeptides have now been identified (Yang et al., 1995; Nagasawa et al., 1999).

This work did not yield any Type II hormones. A different approach was, therefore, adopted, namely to design primers against conserved regions of known Type II hormones which would differentiate them from Type I hormone genes.

PCR of genomic DNA using primers designed on this basis was successful in amplifying a short fragment which, when sequenced, proved to contain a fragment from the coding region of a CHH-family gene with substantial homology to the lobster GIH (De Kleijn et al. 1994). Using this as a probe on an eyestalk cDNA library revealed a cDNA which had not previously been identified. This encoded a Type II CHH-family member with a predicted length for the mature protein of 76 amino acids. Examination of FTMS profiles of sinus glands revealed that a peptide of the predicted molecular weight could be identified. However, the concentrations of this peptide appeared consistently 1-2 orders of magnitude lower than that of the CHH-like molecules. This molecule was designated PmSGPVI.

Although for the reasons discussed above, we could not be confident which sinus gland peptide is the GIH, it was decided to concentrate on PmSGPVI as the Type II hormones include the lobster GIH (De Kleijn et al., 1994) and the crab MOIH (Wainwright et al. 1996). To achieve good protein expression and a ready means of purifying the protein, we expressed the mature PmSGPVI peptide as a fusion to the maltose-binding protein (MBP). Advantages were that the use of MBP stabilized expression of the fusion protein, it allowed ready purification of the fusion protein on an amylose column, and endoproteinases are available to cleave the fusion protein from the MBP carrier, to enable release of the mature neuropeptide (Riggs, 1994).

In initial experiments, the fusion protein was expressed with a signal sequence to direct secretion to the exterior of the cell, to mimic the formation of disulfide bonds that would occur when the neuropeptide is synthesized naturally in prawns. However, yields were disappointingly low and insufficient for antibody production. We then switched to expressing a construct without a signal sequence using a thioredoxin reductase negative strain of *E. coli*, which permits formation of disulfide bonds in the cytoplasm and this resulted in much higher yields (Derman et al. 1993). In one large-scale experiment, we achieved yields of 27 mg/mL of fusion protein, which is approaching the maximum possible yield.

Passive Immunization trials: A diversity of strategies

For initial passive immunization trials, we used the MBP-SGPVI fusion protein to produce polyclonal antisera. Conjugation of small peptides to larger carrier molecules often facilitates development of an appropriate immune response in the host organism and so, in this case, MBP was acting as a carrier protein. The polyclonal antisera were produced in sheep, resulting in very large volumes of antisera being available. The sheep serum was found to contain antibodies that specifically recognized SGPVI after the immunization schedule was complete, indicating that the antibody production was successful (Fig. 7.3). We then further purified the sheep serum to enrich for antibodies that recognize SGPVI. The polyclonal sera is useful for initial proof-of-concept experiments.

We are also in the process of isolating monoclonal antibodies against PmSGPVI. The advantage of monoclonal antibodies is that they offer an unlimited supply of antibody, which could be used for a commercial scale alternative to eyestalk ablation. In addition, monoclonal antibodies can be supplied to animals at higher specific titre, and hence are likely to be more effective.

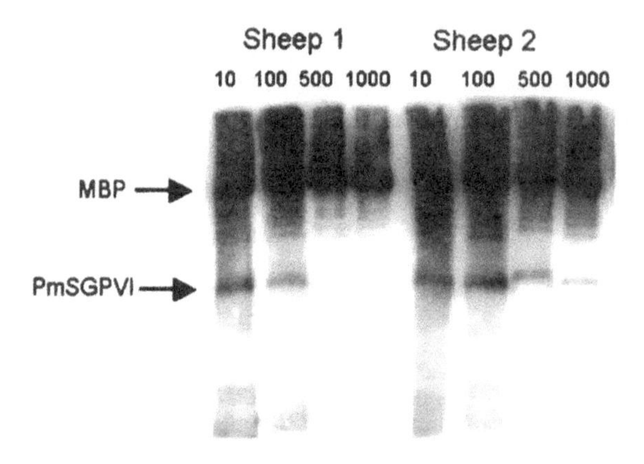

Fig. 7.3. Polyclonal antisera from sheep recognize PmSGPVI. This figure shows antisera used at different dilutions—as indicated above the lanes—to probe a Western blot of MBP-PmSGPVI fusion protein digested with an endoproteinase to separate MBP from the PmSGPVI neuropeptide. Secondary detection was with an anti-sheep-IgG conjugated to horseradish peroxidase (HRP) using a chemiluminescent substrate for HRP. A strong positive reaction to both MBP and PmSGPVI is evident in both sheep, particularly sheep 2. In addition, a number of other proteins from the *E. coli* cultures are recognized by the polyclonal sera.

A further hurdle to be overcome is to establish the optimal means of delivery of antibodies to prawns. There are reports of successful use of antisera to immunize *P. monodon* against bacterial infection. However, in this study, the longevity of the antibodies in the haemolymph was not evaluated (Lee et al. 1997). Our results indicate that mouse or sheep IgGs injected into prawn tail muscle have a half-life of 12 hours or less for detectable persistence in the hemolymph. Clearly, this could be a problem if persistent removal of GIH from the haemolymph is required to mimic the effects of eyestalk ablation.

Possible reasons for the short half-life of mammalian IgG in prawns are either specific degradation or leaching. We have carried out in vitro studies which have demonstrated that the short half-life is not due to endogenous proteases in prawn blood. Rapid leaching can only be controlled by continually infusing mammalian IgG into the circulatory system to ensure that sinus gland peptides are effectively neutralized. A slow release depot of mammalian IgG is under development to ensure continual infusion of anti-GIH IgG into the haemolymph.

An alternative strategy to passive immunization is the use of inhibitory RNA or RNAi (Sharp, 1999; Bass, 2000; Bosher and Labouesse, 2000). In this technique, double-stranded RNA (dsRNA) corresponding to the sequence of a target gene is injected into the animal, and results in effectively switching off production of any protein encoded by that gene. As we had full length cDNA available, we used this to prepare stocks of dsRNA corresponding to the SGPVI gene to incorporate into broodstock trials. If successful, this RNAi should inhibit production of GIH and, hence, lead to ovarian development once the circulating and stored levels of GIH have been depleted.

We carried out an initial trial of the passive immunization and RNAi techniques with promising results. However, this trial was on a very small scale and such trials need to be expanded to rigorously demonstrate whether spawning induction is actually working. If an expanded "proof-of concept" trial is successful, extensive further work will be required to develop optimal dose rates and methods of delivery. It will also be important to trial monoclonal antibodies, as opposed to polyclonal antibodies, when they become available.

Thus, much work remains to be done. However, we are optimistic that at the conclusion of this project, we will have achieved a more reliable and specific means of inducing spawning in both wild and domesticated broodstock. We will also have made progress in methods to deliver molecules to prawns, including both mammalian antibodies and dsRNA, which may be of use to the prawn industry in other roles, such as in the fight against disease.

ROCK LOBSTER AQUACULTURE RESEARCH

World Market

World landings of rock lobsters are dominated by the *Astacidea*, or clawed lobsters, from the north Atlantic. The *Palinuridea*, or rock lobsters (also referred to as spiny lobsters), make up approximately 37% of world landings. Of these, tropical species (including the Caribbean lobster) make up 70% of rock lobster landings (Fig. 7.4). As clawed lobsters are not an Australian species, they are not candidates for aquaculture in Australia.

Rock lobsters account for 20% of the total value of fisheries production in Australia, and are the single most valuable fisheries commodity (ABARE, 2002).

However, the wild harvest has now plateaued and the only possible means to expand production is by development of an aquaculture industry.

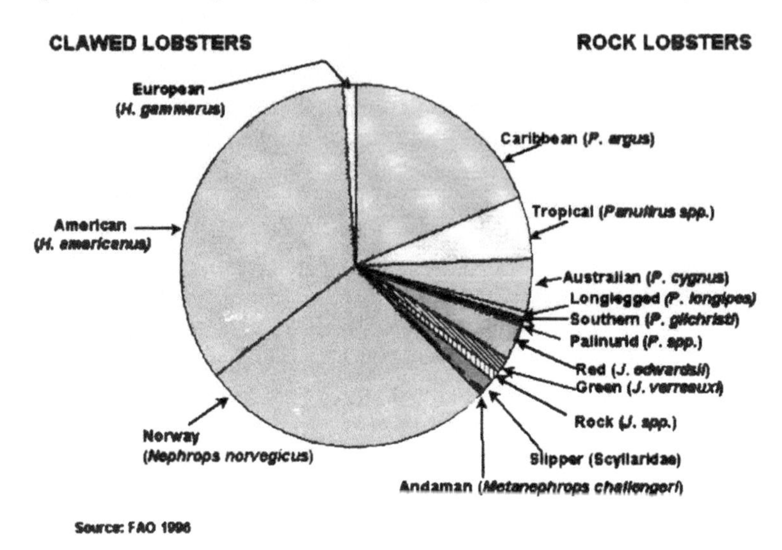

Fig. 7.4: Annual world harvest of clawed and rock lobsters. Rock lobsters make up 37% of the world catch of 210,000 metric tonnes. Tropical rock lobsters comprise 25% of the world catch and 70% of the rock lobster catch.

Species selection

Attempts at the aquaculture production of tropical rock lobsters (Palinuridea) span over a quarter of a century (Cobb and Phillips 1980; Kittaka 1997). Major research efforts have taken place in Japan, New Zealand and the USA (Moe 1991; Kittaka 1997). None of these efforts have met with commercial success. The major hurdle to be overcome for the aquaculture production of rock lobsters is that of larval rearing. The rock lobster has one of the longest planktonic larval phases of any marine organism. The planktonic larval phase is complex and characterized by delicate, 'spider-like' larvae known as phyllosomas (Fig. 7.5). Typically, there are 11 morphological phases—with even more moults—that the larvae pass through before metamorphosing into pueruli (and then into juvenile lobster) (Booth and Phillips, 1994; Kittaka and Abrunhosa, 1997).

Because of the difficulties encountered in rearing rock lobster larvae, a considerable amount of effort has been directed towards the collection of newly-settled juveniles from the wild, in order to bypass the larval rearing phase of the production cycle. This is, in fact, the basis of a successful cottage industry in coastal villages in Vietnam, where such juveniles are on-grown to about 1 kg in cages in coastal bays. However, the unpredictability of supply due to high variation in annual recruitment of juveniles—coupled with questions about sustainability of an industry reliant on collection of wild juveniles—makes this a difficult prospect for a large scale commercial industry.

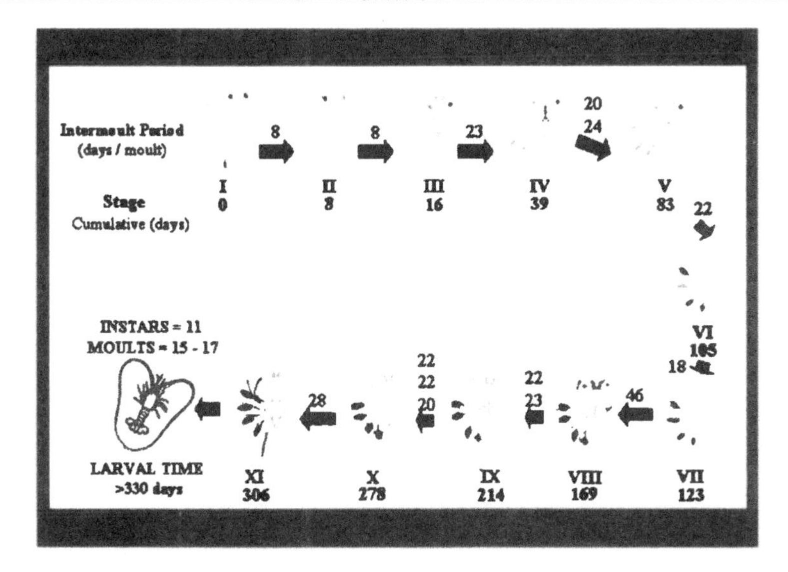

Fig. 7.5: An example of the complex larval phase of rock lobsters. This is the larval phase of a temperate South African species, *Jasus lalandii.* Each stage requires a moult before change into the next stage. Hormones trigger each moult.

Hence, if rock lobster aquaculture is going to develop into a viable sustainable industry, the closing of the life cycle is necessary. The selection of a particular species as an aquaculture candidate is determined by the possession of favourable characteristics that are amenable to farming conditions. Amongst others, these include suitable life history traits, ability to be reared in high densities, disease resistance, nutritional requirements, and tolerance of variations in water quality.

The only commercially viable crustacean aquaculture industries at present are those which farm species with a short larval phase (Lee and Wickens, 1992). The larval rearing period is considered to be the most labour intensive and technically difficult, and is often associated with high mortality, which may be due to pathogenic microbial infections. By far, the largest crustacean aquaculture sector is that for penaeid prawns. Penaeids have a larval phase of only 15-20 days. Success in larval rearing of some crabs has also been reported and in all the cases, they also have a short (15-30 day) larval life. Thus, it can be concluded that length of the larval phase is a key criterion for selecting candidate rock lobster species for aquaculture.

Among rock lobster species, some tropical species appear to have a relatively short larval phase, around 120 to 200 days, such as *P. ornatus* (ornate rock lobster) as shown in Fig. 7.6. The Australian tropical *P. homarus* (scalloped rock lobster), *P. polyphagus* (mud spiny lobster), and *P. longpipes* (long-legged spiny lobster) species may also have relatively short larval phases. These may all be suitable aquaculture candidates, as they also appear to be tolerant of highly variable water quality.

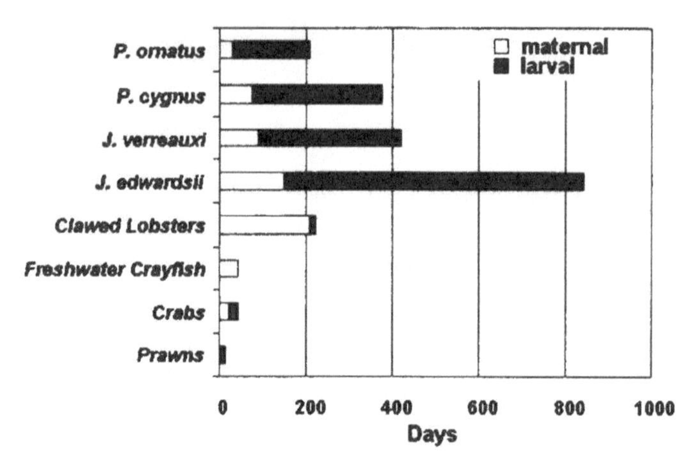

Fig. 7.6: Tropical lobsters (e.g. *Panulirus ornatus*) have the shortest maternal and larval phases of Australian rock lobsters.

Hormonal manipulation of larval development

The key to achieving aquaculture of rock lobster will be improving larval survival. One potential breakthrough could be to shorten the larval phase. Temperature and photoperiod regimes can modify larval development (Mikami and Greenwood, 1997). However, such regimes can lead to high mortalities of larvae rather than improvements in quality or rate of development. In any case, there has been no dramatic shortening of the larval phase of any crustacean by using this approach. There is an opportunity to potentially and significantly shorten the larval life by hormonal manipulation.

Larval transformations in insects, the nearest relatives to Crustacea, are regulated by hormonal cycles, primarily changes in ecdysteroid titres. Likewise, the various transformations through the larval stages and moults are under endocrinological and molecular controls in Crustacea (Chang, 1984). All the major endocrine organs, including the X-organ-sinus gland complex in the eyestalk, the Y-organ and the mandibular organ are involved. The X-organ-sinus gland complex is the source of moult-inhibiting hormone (MIH), as discussed earlier for prawns. The Y-organ is the source of ecdysteroids which are the major steroid hormones-regulating moult. The mandibular organ is the source of methyl farnesoate (MF), a juvenile-like hormone, which is believed to be involved in modulating other hormone actions in both larvae and adults.

In Astacid lobsters, episodic pulses of ecdysteroids are correlated with larval transformations (Chang, 1984). Steroids are known to function as critical regulators in insect metamorphosis (Truman and Riddiford, 1999). Ecdysteroids act by binding to receptors in the nucleus and inducing conformational changes in those receptors which lead to the activation or inhibition of specific gene expression. Much of the temporal and tissue specificity of ecdysteroid responsiveness is modulated by the presence or absence of specific nuclear receptors (Perera et al. 1998). In a majority of the cases, the receptor functions as a dimer—either a homodimer or a heterodimer—with a different nuclear hormone receptor. For instance, in *Drosophila*, a receptor named ultraspiricle (USP) will form a heterodimer with the ecdysone receptor (EcR) and this stabilizes the binding of the ecdysone receptor to DNA in the presence of 20-OH ecdysone (Jones and Sharp, 1997). Alternatively, a third receptor termed DHR38 can bind to USP and sequester it, to prevent activation of USP/EcR dependent genes (Sutherland et al. 1995).

The overall goal of the current AIMS project is to analyse hormonal inputs into rock lobster larval development and determine whether provision of exogenous hormones can be used to manipulate the length of the larval phase.

The family of nuclear receptors involved is highly conserved and a pair of such hormone receptors has been cloned from the fiddler crab, the first such receptor genes to be isolated from a crustacean (Chung et al. 1998). Due to the conservation of sequence, it should prove possible to clone the corresponding homologues from the lobster *P. ornatus* and to develop specific PCR probes for each nuclear receptor. Changes in the abundance of the messenger RNAs specifying these receptors can then be assayed using specific, quantitative PCR.

In parallel with the molecular biology experiments, high performance liquid chromatography (HPLC) assays for the measurement of MF and various ecdysteroids will be used to examine the temporal changes in hormone titres in groups of phyllosomas. Together, this will give an overall picture of how changes in hormone levels might be translated into changes in gene expression. In particular, it will be of interest to examine changes in phyllosoma that are simply undergoing a moult within a stage compared to those which are undergoing a metamorphic transition between stages.

REFERENCES

ABARE. Australian Fisheries Statistics, 2002. 2003. Canberra, Australia. pp. 69.

Argue, B.J., Arce, S.M., Lotz, J.M., and Moss, S.M. 2002. Selective breeding of Pacific white shrimp (*Litopenaeus vannamei*) for growth and resistance to Taura syndrome virus. Aquaculture, 204: 447-460.

Bass, B.L. 2000. Double-stranded RNA as a template for gene silencing. Cell, 101: 235-238.

Booth, J.D. and Phillips, B.F. 1994. Early life history of spiny lobster. Crustaceana, 66: 271-294.

Bosher, J.M. and Labouesse, M. 2000. RNA interference: genetic wand and genetic watchdog. Nature Cell Biol., 2: E31-E36.

Chang, E.S. 1984. Ecdysteroids in Crustacea: Role in reproduction, molting and larval development. In: Advances in Invertebrate Reproduction, (eds) Engels, W. Clark, W.H., Fischer, A. Olive, P.J.W. and Went, D.F. Elsevier Science Publishers, Amsterdam, pp. 223-230.

Chang, E.S., Chang, S.A., Beltz, B.S., and Kravitz, E.A. 1999. Crustacean hyperglycemic hormone in the lobster nervous system: Localization and release from cells in the subesophageal ganglion and thoracic second roots. J. Comp. Neurol. 414: 50-56.

Chung, A.C., Durica, D.S., and Hopkins, P.M. 1998. Tissue-specific patterns and steady-state concentrations of ecdysteroid receptor and retinoid-X-receptor mRNA during the molt cycle of the fiddler crab, *Uca pugilator*. Gen. Comp. Endocrinol., 109: 375-389.

Chung, J.S., Dircksen, H., and Webster, G. 1999. A remarkable, precisely timed release of hyperglycaemic hormone from endocrine cells in the gut is associated with ecdysis in the crab *Carcinus maenas*. Proc. Natl. Acad. Sci. USA, 96: 13103-13107.

Cobb, J.S., and Phillips, B.F. 1980. The Biology and Management of Lobsters. Academic Press, New York.

Crocos, P.J., Preston, N.P., Smith, D.M, and Smith, M.R. 1999. Reproductive performance of domesticated *Penaeus monodon*. Abstracts, World Aquaculture 99, Sydney, p. 185.

Davey, M.L., Hall, M.H., Willis, R.H., Oliver, R.W.A., Thurn, M.J., and Wilson, K.J. 2000. Five crustacean hyperglycemic family hormones of *Penaeus monodon*: Complementary DNA sequence and identification in single sinus glands by electrospray ionization-Fourier transform mass spectrometry. Mar. Biotechnol., 2: 80-91.

De Kleijn, D.P., Sleutels, F.J., Martens, G.J., and Van Herp, F. 1994. Cloning and expression of mRNA encoding prepro-gonad-inhibiting hormone (GIH) in the lobster *Homarus americanus*. FEBS Letters, 353: 255-258.

De Kleijn, D.P., de Leeuw, E.P., van den Berg, M.C., Martens, G.J., and Van Herp, F. 1995. Cloning and expression of two mRNAs encoding structurally different crustacean hyperglycemic hormone precursors in the lobster *Homarus americanus*. Biochim. Biophysic. Acta, 1260: 62-66.

Derman, A.I., Prinz, W.A., Belin, D., and Beckwith, J. 1993. Mutations that allow disulfide bond formation in the cytoplasm of *Escherichia coli*. Science, 262: 1744-1747.

Dircksen, H., Webster, S.G., and Keller, R. 1988. Immunocytochemical demonstration of the neurosecretory systems containing putative moult-inhibiting hormone and hyperglycaemic hormone in the eyestalk of brachyuran crustaceans. Cell Tissue Res., 251: 3-12.

Dircksen, H., Boecking, D., Heyn, U., Mandel, C, Chung, J.S., Baggerman, G., Verhaert, P., Daufeldt, S., Ploesch, T., Jaros, P.P., Waelkens, E., Keller, R. and Webster, S.G. 2001. Crustacean hyperglycaemic hormone (CHH)-like peptides and CHH-precursor-related peptides from pericardial organ neurosecretory cells in the shore crab, *Carcinus maenas*, are putatively spliced and modified products of multiple genes. Biochem. J., 356: 159-170.

FAO 1996. Fishery Statistics: Capture Production, Vol. 82, Food and Agriculture Organisation of the United Nations, Rome. pp. 365-370.

FAO 2003. Review of the state of world aquaculture. Food and Agriculture Organisation of the United Nations, Rome. p. 95.

Fingerman, M. 1992. Glands and secretion. In: Microscopic Anatomy of Invertebrates, (eds) Harrison, F.W. and Humes, A.G., John Wiley, New York, pp. 345-394.

Goyard, E., Patrois, J., Peignon, J.M., Vanaa, V., Dufour R., Viallon, J. and Bédier, E. 2002. Selection for better growth of *Penaeus stylirostris* in Tahiti and New Caledonia. Aquaculture, 204: 461-468.

Jones, G. and Sharp, P.A. 1997. Ultraspiracle: An invertebrate nuclear receptor for juvenile hormones. Proc. Natl. Acad. Sci. USA, 94: 13499-13503.

Keller, R. 1992. Crustacean neuropeptides: structure, functions and comparative aspects. Experientia, 48: 439-448.

Kenway, M.J. and Benzie, J.A.H. 1999. High quality larvae from tank-reared *Penaeus monodon*. Abstracts, World Aquaculture 99, Sydney, p. 388.

Khayat, M., Yang, W., Aida, K., Nagasawa, H., Tietz, A., Funkenstein, B., and Lubzens, E. 1998. Hyperglycaemic hormones inhibit protein and mRNA synthesis in in vitro incubated ovarian fragments of the marine shrimp *Penaeus semisulcatus*. Gen. Comp. Endocrinol., 110: 307-318.

Kittaka, J. 1997. Application of ecosystem culture method for complete development of phyllosomas of spiny lobster. Aquaculture, 155: 319-331.

Kittaka, J., and Abrunhosa, F.A. 1997. Characteristics of palinurids in larval culture. Hydrobiologia, 358: 305-311.

Lacombe, C., Greve, P., and Martin, G. 1999. Overview on the sub-grouping of the crustacean hyperglycemic hormone family. Neuropeptides, 33: 71-80.

Laverdure, A.M., Carette-Desmoucelles, C., Breuzet, M., and Descamps, M. 1994. Neuropeptides and related nucleic acid sequences detected in penaeid shrimps by immunohistochemistry and molecular hybridizations. Neuroscience, 60: 569-579.

Lee, D.O.C. and Wickens, J.F. 1992. *Crustacean Farming*. Oxford: Blackwell Scientific Publications, Oxford.

Lee, K.-K., Liu, P.C., Kou, G.H., and Chen, S.-N. 1997. Passive immunisation of the tiger prawn, *Penaeus monodon*, using rabbit antisera to *Vibrio harveyi*. Lett. Appl. Microbiol. 25: 34-37.

Liu, L. and Laufer, H. 1996. Isolation and characterization of sinus gland neuropeptides with both mandibular organ inhibiting and hyperglycemic effects from the spider crab *Libinia emarginata*. Arch. Insect Biochem. Physiol. 32: 375-385.

McIntosh, R.P. 2000. Changing paradigms in shrimp farming: III. Pond design and operation considerations. Global Aquacult. Advocate, 3: 42-45.

Menasveta, P., Sangpradub, S., and Piyatirativorakul, S. 1994. Effects of broodstock size and source on ovarian maturation and spawning of *Penaeus monodon* Fabricius from the Gulf of Thailand. J. World Aquacult. Soc., 25: 141-149.

Mikami, S. and Greenwood, J.G. 1997. Influence of light regimes on phyllosomal growth and timing of moulting in *Thenus orientalis*. Mar. Freshwater Res., 48: 777-782.

Moe, M.A. 1991. *Lobster: Florida, Bahamas, the Caribbean.*, Green Turtle Publications. Plantation, Florida. pp. 239–302.

Nagasawa, H., Yang, W.J., Aida, K., and Sonobe, H. 1999. Chemical and biological characterisation of neruopeptides in the sinus glands of the kuruma prawn, *Penaeus japonicus*. In: Peptide Science—Present and Future, (ed.) Shimonishi, Y. Kluwer Academic Publishers, Amsterdam, pp. 453-454.

O'Sullivan, D. and Dobson, J. 2000. Status of Australian Aquaculture in 1998/99. Austasia Aquacult. Trade Direct. 2000/2001, 3-16.

Ohira, T., Watanabe, T., and Nagasaw, H. 1997. Cloning and sequence analysis of a cDNA encoding a crustacean hyperglycemic hormone from the kuruma prawn *Penaeus japonicus*. Mol. Mar. Biol. Biotechnol., 6: 59-63.

Perera, S.C., Palli, S.R., Ladd, T.R., Krell, P.J., and Retnakaran, A. 1998. The ultraspiracle gene of the spruce budworm, *Choristoneura fumiferana*: Cloning of cDNA and developmental expression of mRNA. Dev. Genet., 22: 169-179.

Preston, N.P., Brennan, D.C., and Crocos, P.J. 1999. Comparative costs of post-larval production from wild or domesticated *Penaeus japonicus* broodstock. Aquacult. Res. 30, 191-197.

Riggs, P. 1994. Expression and purification of maltose-binding protein fusions. In: *Current Protocols in Molecular Biology*, (eds) Ausubel, F.M., Brent, R., Kingston, R.E., Moore, D.D., Seidman, J.G., Smith, J.A. and Struhl, K., John Wiley, New York. pp. 166: 1-14.

Rosenberry, B. 2003. World Shrimp Farming 2003, Shrimp News International, San Diego, CA.

Sharp, P.A. 1999. RNA and double-strand RNA. Genes Dev. 13: 139-141.

Soyez, D. 1997. Occurrence and diversity of neuropeptides from the crustacean hyperglycaemic hormone family in arthropods: A short review. Annal. New York Acad. Sci., 814: 319-323.

Sutherland, J.D., Kozlova, T., Tzertzinis, G., and Kafatos, F.C. 1995. Drosophila hormone receptor 38: A second partner for *Drosophila* USP suggests an unexpected role for nuclear receptors of the nerve growth factor-induced protein B type. Proc. Natl. Acad. Sci. USA, 92: 7966-7970.

Truman, J.W. and Riddiford, L.M. 1999. The origins of insect metamorphosis. Nature, *401: 447-452*.

Wainwright, G., Webster, S.G., Wilkinson, M.C., Chung, J.S., and Rees, H.H. 1996. Structure and significance of mandibular organ-inhibiting hormone in the crab, *Cancer pagurus*. involvement in multihormonal regulation of growth and reproduction. J. Biol. Chem. 271: 12749-12754.

Webster, S.G. 1998. Neuropeptides inhibiting growth and reproduction in crustaceans. In: Recent Advances in Arthropod Endocrinology, (eds) Coast, G.M. and Webster, S.G., Cambridge University Press, pp. 33-52.

Yang, W., Aida, K., and Nagasawa, H. 1995. Amino acid sequences of a hyperglycaemic hormone and its related peptides from the kuruma prawn, *Penaeus japonicus*. Aquaculture 135: 205-212.

Chapter 8

Homeopathic Induction of Spawning in Ornamental Fish

R.Vishakan, S.Balamurugan, C.Maruthanayagam and **P.Subramanian**
Department of Animal Science, Bharathidasan University
Tiruchirappalli - 620 024, Tamilnadu, India
Email: visak123@rediffmail.com

ABSTRACT

The present study investigates a homeopathic preparation for induced breeding ornamental fish. Intramuscular injection of *Natrum muriaticum* induces a female goldfish to spawn within 24 hours, while the control animal took 5 days to spawn.

Keywords: Goldfish, *Natrum muriaticum*, injection

INTRODUCTION

In aquaculture industry, ornamental fish have gained a great deal of attraction, and its demand has increased in recent years. The development of this segment of the aquaculture industry depends on virtually the seed availability in nature as well as in hatchery. Hatchery seeds have many advantages over the wild ones. To induce breeding with hormones, either a gonadotropin (fish or mammalian, pure or crude) or a GnRH analogue (GnRHa) with or without a dopamine antagonist are used (Harvey and Carolsfeld, 1993).

Since the first successful attempt of induced breeding of *Cresterdon decemmaculatus* by injection of pituitary extract (Houssay, 1931), the search for newer and better methods for induction of spawning in teleost fishes has gained momentum. Mammalian gonadotropin (LH, FSH, HCG and PMSG) and adrenocorticoid steroids have been used to spawn a variety of fishes and come out with variable results (Homeswar et al. 1992). For hatchery production, the pituitary gland, HCG and ovaprim injection have been found effective. The use of oxytocin, synahorin and prolan had little success. Many salts and herbs were suggested from Indian medicines to improve the fertility and reproductive ability of humans. One such homeopathic preparation is *Natrum muriaticum* (Satyaprakash, 1986), but it has not so far been used except by Mitra and Raizada (1986) who used this preparation to induce ovulation in food fishes *Labeo rohita* and *Cirrhina mrigala*. This preparation was also used to induce breeding goldfish *Carassius auratus*.

MATERIAL AND METHODS

Sexually mature male and female goldfish with body weight of 250 g were procured from the local aquarist and brought to the laboratory in polythene bags. They were acclimated to laboratory conditions for 15 days in 40-l glass aquaria, simulating the natural habitat of the fish. The temperature was maintained at 27. 2 ± 2°C with normal diffused illumination.

The homeopathic preparation *Natrum muriaticum* of hundred thousand potency (CM) was prepared and administered to the females at a dose of 1 ml/kg body weight. The females were injected intramuscularly below the dorsal fin. For control animal, the physiological saline was injected. The changes in the body weight of fish before and after the treatment were recorded.

RESULTS AND DISCUSSION

Injection of the homeopathic preparation *Natrum muriaticum* into the goldfish induced the spawning within 22 hours against 5 days in the control (Table 8.1). The experimental animals laid 2,856 eggs while the control laid 209 eggs. The ovulation induced by injections or implants of luteinizing hormone releasing hormone analogues in goldfish is relatively ineffective but with the combination of pimozide, a dopamine antagonist, ovulation was effective (Sokolowska et al., 1984). HCG (450 IU/kg each) combined with pituitary of *Tachysurus sp* caused spawning in river catfish *Pangassius sutchi* and Thai catfish *Clarias macrocephalus* after only two doses (Saidin et al. 1988). Conversely, homeopathic preparation induced spawning within 24 hr and even with a single dose. In the Indian carp *Cirrhina mrigala* (HAM), a single dose administeration of LH-RH analogue (10 μg/kg) and pimozide (10 mg/kg) caused ovulation (Kaul et al., 1986). The cost effectiveness of *Natrum Muriaticum* compares favourable with HCG, GnRHa and pituitary extracts.

Table 8.1: Induced spawning of goldfish administrated with *Natrum muriaticum*

Fish	*Body weight (g)*	*Eggs spawned (no.)*	*Latency periods (hr)*
Control	253 ± 1.8	209 ± 21	120 ± 0.30
Experimental	250 ± 1.3	2856 ± 16	22 ± 1.0 hours

ACKNOWLEDGEMENTS

The authors wish to thank the authorities of Bharathidasan University for providing facilities.

REFERENCES

Harvey, B. and Carolsfeld, J. 1993. Induced Breeding In Tropical Fish Culture. IDRC, Ottawa, pp. 144.

Homeshwar, M. and Singh. H. 1992. Induced ovulation of rice field eel (*Monopterus albus*) by a synthetic non-peptide. Indian J. Exp.Biol., 30: 1111-1113.

Houssay, B.A. 1931. Action sexualle de l'hypophyse sur les poissons et les repetiles. C.R. Soc. Biol., Paris, 106: 377-378.

Kaul, M. and Rishi, K.K. 1986. Induced spawning of the Indian major carp, *Cirrhina mrigala* (HAM) with LH-RH analogue or pimozide. Aquaculture, 54: 45-48.

Mitra, S.D. and Raizada, S. 1986. Gonadal development of major carps through homeopathy. Fishing Chimes, May, 13: 17.

Saidin, T. Othman, A.A. and Sulaiman, M.Z. 1988. Induced spawning techniques practiced at *Batu Berendam*, Melaka. Aquaculture, 74: 23-33.

Satyaprakash, 1986. The homeopathic medicine, Rational Life Publications, New Delhi, Volume 1, p. 522.

Sokolowska, M., Peter, R.E., Nahorniak, C.S., Pan, C.H., Chang, J.P., Crim, L.W. and Weil, C. 1984. Induction of ovulation in Gold fish, *Carassius auratus*, by pimozide and analogues of LH-RH. Aquaculture, 36: 71-83.

Chapter 9

Asymptomatic Mortality of *Penaeus indicus* Infected by Non-luminescent *Vibrio harveyi*

E. S. Abdel-Aziz * and **S.M. Dass** **
* Department of Fish Diseases, Animal Health Research Institute
Agricultural Research Centre, Dokki, Giza, Egypt
** Department of Aquaculture, Ministry of Agriculture and Water
Riyadh, Kingdom of Saudi Arabia
Email: drsmdas@hotmail.com

ABSTRACT

Asymptomatic mass mortalities were recorded in the summer months among cultured *Penaeus indicus* broodstock and post-larvae (PL15). Two non-luminescent *Vibrio harveyi* biotypes (*V.h.1* and *V.h.2*) were dominantly identified along with other bacterial species, mainly *Pseudomonas fluorescens (P.fl.), Aeromonas hydrophila (A.h.), Staphylococcus* species (*S. spp.*). Incidental occurrence of *V.h.1, V.h.2, P.fl., A.h.* and *S. spp.* frequency in infected samples peaked 82.0, 45.3, 8.0, 4.7 and 7.3%, respectively. The disease was probably linked to the adverse environmental conditions like high levels of temperature, salinity, ammonia and nitrite in infected ponds and the involvement of highly virulent *V. harveyi* biotypes. The combination of these factors revealed in the development of a peracute course of infection were characterized by high mortalities and almost absence of clinical signs. The recovered *V. harveyi* biotypes were sensitive to oxytetracycline, chloramphenicol, enrofloxacin, nalidixic acid and oxolinic acid, while resistant to ampicillin, novobiocin, kanamycin, gentamycin and sulfonamide.

Keywords: Vibriosis, environmental stress, antibiotic sensitivity, immuno compromise.

INTRODUCTION

Vibriosis, a disease caused by different *Vibrio* species, is one of the major threats affecting fish and shellfish mariculture worldwide (Lightner, 1988; Austin and Austin, 1993; Hjeltnes and Roberts, 1993). One of these species, *Vibrio harveyi* occurs naturally in warm marine waters , as also on the surface and in the gut of marine aquatic animals (Ruby and Morin, 1979). The occurrence of vibriosis in penaeid shrimps is generally influenced by factors such as stress, environmental contamination, and high counts of potentially pathogenic bacteria (Chen, 1992; Nash et al., 1992; Mohaney et al., 1994; Ruangpan et al., 1995). High mortality of larvae and juveniles of the tiger prawn (*P. monodon*) associated with luminiscent *V. harveyi* has been observed in hatcheries or farms in India (Karunasagar et al., 1994), Thailand (Jiravonichpaisal et al., 1994), Australia (Pizzutto and Hirst, 1995), Taiwan (Chen et al., 1992), Indonesia (Sunaryanto and Mariam, 1986) and Philippines (Baticados et al. 1990).

Presence of white spots in the carapace was the major clinical finding in recent outbreaks caused by *Vibrio* species in cultured peneid shrimps (Takahashi et al., 1994; Yu, 1995). However, outbreaks caused by *V. harveyi* among cultured *P. monodon* without gross signs are also recorded (Liu et al. 1996). The use of antibiotics has been shown to be quite effective in controlling vibriosis in laboratory experimental trials (Baticados et al., 1990; Rukyani et al., 1992). However, the expected results were not realized when applied in fields (Chanratchkool et al., 1995), where long term or improper usage of antibiotics might have resulted in the development of antibiotic-resistant strains.

This chapter deals with the non-luminescent prevalence, characteristics, and antibiotic sensitivity of isolated *V. harveyi* from cultured *P. indicus* broodstock and post-larvae (PL 15) as well as from rearing pond waters after mass mortalities without characteristic gross signs of infection were observed.

MATERIAL AND METHODS

Sampling and Observations

Shrimp samples : Moribund and freshly died *P. indicus* broodstock and post-larvae (PL 15) were collected from commercial shrimp farms on the Red Sea coast, where mass mortalities were observed. Samples were clinically examined for gross signs and lesions indicating disease conditions by way of the methods described by Austin and Austin (1989).

Water samples: Pond water samples were collected from the affected farms and analysed for various physico-chemical parameters (dissolved oxygen, salinity, pH, temperature, ammonia and nitrite) by using the standard analytical methods (Dass, 1989).

Fifty infected specimens and 10 control specimens of broodstock were dissected for the extraction of bacteriological samples. The samples were aseptically extracted from the hepatopancreas and streaked on nutrient agar (Difco), Mac Conkey agar (Difco) and Tryptic Soy agar (oxoid, supplemented with 2% NaCl) Austin and Austin (1989).

Fifty infected postlarval samples and 10 control *P. indicus* were washed three times with sterile seawater, placed in sterile test tubes containing 2 ml sterile seawater and macerated with sterile glass rods. In addition, water samples were taken from fifty affected ponds. Post-larvae and water samples were cultured using the same media mentioned above. All plates were incubated at room temperature 25°C for 24 hr. The recovered isolates were identified using standard morphological, physiological, and biochemical characteristics Krieg and Holt (1984), Lennette et al., (1985) as per the system of Bio Merieux, France API 20 E.

SENSITIVITY OF *V. HARVEYI* BIOTYPES TO ANTIBIOTICS

Recovered *V. harveyi* biotypes were grown on TSA (with 2% NaCl) for 24 hr at 25°C. The bacteria were suspended in sterile PBS and diluted to a turbidity equivalent to Mac Farland No. 0.5 standard solution (0.5 ml BaS04 + 99.5 ml 0.36N Hcl). A volume of 0.2 ml of each bacterial suspension was spread onto Muller - Hinton agar (Difco) and then antibiotic discs were added (Koneman et al., 1988). The antibiotic discs (Bio Merieux, France) contained oxytetracyline, chloramphenical, ampicillin, novobiocin, oxolinic acid, nalidixic acid, enrofloxacin, kanamycin, gentamycin and sulphonamide. The plates were incubated for 18 hours at 25°C and the inhibition zones were measured and compared to the standards, of Bio Merieux, France.

Mycological (Austin and Austin, 1989; Roberts,1989) and parasitological (Duijn, 1973; Post, 1987) examinations of the collected samples were carried out.

RESULTS AND DISCUSSION

The rapid expansion of intensive shrimp culture and the upgradation of traditional shrimp farms worldwide have met with increased incidences of diseases and epizootics. The predominance of infections with *Vibrio* species in shrimp culture (Austin and Austin, 1993) has created the need to understand the epizootiology, pathogenesis and mechanisms by which this disease occurs that might be of significance in disease prevention and control programs.

As hepatopancreas of shrimp is reportedly the main target organ of most bacterial pathogens (Chen, 1992), analysis of the control was restricted to hepatopanceas. Bacteriological examination of the samples revealed the presence of non-luminescent *V. harveyi* isolates from hepatopancreas of broodstock, the entire body of post-larvae (15) and in rearing water samples, along with other bacterial species (Table 9.1).

Majority of broodstock and post-larvae (15) samples showed no postmortem changes. However, a few broodstock shrimp showed pathological alterations in the hepatopancreas in the form of paleness, haemorrhagic patches and steaks, and textures soft, friable, or liquified. Significant symptomatic mortality in cultured larvae caused by luminescent *V. harveyi* has been documented (Prayitno and Latchford, 1995). On the other hand, Liu et al., (1996) and Lavilla-Pitogo et al., (1998) recorded the occurrence of asymptomatic disease outbreaks with high mortality among cultured kuruma prawn (*P. japonicus)* and tiger shrimp (*P. monodon)* caused by luminescent *V. harveyi.*

Table 9.1: Recovery (%) of bacterial isolates from shrimp and water samples.

Sample source	*Nature of sample*	*V.h.1*	*V.h.2*	*P.fl.*	*A.h.*	*S. spp.*
Hepatopancreas	Control	Nil	Nil	Nil	Nil	Nil
Broodstock	Infected	82	14	2	2	Nil
Post-larvae (15)	Infected	70	44	6	4	6
Mascerated whole body	Control	50	30	20	10	20
	Infected	94	78	14	10	16
Pond water	Control	60	50	20	10	20
	Infected	82	45	8	5	7

The recovered bacterial species of clinical importance were *V.h., P.fl., A.h.,* and *S.spp.* Two biotypes of *V. harveyi* (*V.h.*1 and *V.h.*2) were found. The incidence of recovery of *V. h.*1, *V.h.*2, *P.fl. A.h.* and *S. spp.* from infected samples peaked to 82.0, 45.3, 8.0, 4.7 and 7.3%, respectively, whereas these values were 36.7, 26.7, 13.3, 6.7 and 13.3% in control samples, respectively. The higher incidence and prevalence of *V. harveyi* biotypes 1 and 2 may reflect a degree of host specificity and/or environmental selection for the survival and domination of *V. harveyi* in mariculture elements, due to its virulence and competitive elimination of other bacterial species. The domination of the biotype 1 over biotype 2 may be related to its virulence characteristics, and tissue tropism evoked during host-pathogen interaction (Bullock et al. 1971).

The recovery of *V. harveyi* biotypes from the hepatopancreas of broodstock but not from control samples (Table 9.1) may suggest the involvement of highly virulent *V. harveyi* strains rather than the occurrence of a carrier state or latent form of infection. However, Leano et al., (1998) recorded consistent recovery of luminescent vibrios from pond-reared *P. monodon* juveniles with and without disease signs. On the other hand, Lavilla-Pitigo et al. (1990) recorded the isolation of luminescent *V. harveyi* from infected *P. monodon* larvae and rearing water but not from uninfected ones. Meanwhile, the recovery of *V. harveyi* from control post larval (PL 15) samples in this study could be due to the utilization of macerated whole bodies of moribund post-larvae for bacteriological examination since *Vibrio* species naturally occur on the surface and in the gut (Ruby and Morin, 1979).

Table 9.2: Biochemical and growth characteristics of recovered Gram-negative bacterial isolates.

Test	*Recovered isolate*			
	V. h.1	*V.h.2*	*P.fl.*	*A.h.*
Motility	+	+	+	+
β-galactosidase (ONPG)	+	+	-	+
Arginine dihydrolase (ADH)	-	-	+	+
Lysine decarboxylase (LDC)	+	+	-	-
Ornithine decarboxylase (ODC)	+	+	-	-
Tryptophane deaminase (TDA)	-	-	-	-
Gelatinase (GEL)	+	-	-	+
Cytochrome oxidase (OX)	+	+	+	+
Urease (URE)	-	+	-	-
H2S (H2S)	-	-	-	-
Indole (IND)	+	-	-	+
Acetoin (VP)	-	-	+	-
Citrate (CIT)	-	+	+	+
Glucose (GLU)	+	+	-	+
Mannitol (MAN)	+	-	-	+
Inositol (INO)	-	-	-	-
Sorbitol (SOR)	-	-	-	-
Rhamnose (RHA)	-	-	-	-
Sucrose (SAC)	+	-	-	+
Melibiose (MEL)	-	-	-	-
Amygdalin (AMY)	-	-	-	-
Arabinose (ARA)	-	-	-	-
Differential characteristics :				
Luminescence	-	-	-	-
Diffusible greenish yellow pigment production	-	-	+	-
Sensitivity to 0/129 (150 mg)	+	+	-	-
Growth on nutrient agar	-	-	+	+
Growth on TSA + 2 % NaCl	+	+	+	+
Growth on TSA + 7 % NaCl	+	+	-	-
Growth on Cetramide	-	-	+	-
Growth at 4°C	-	-	+	-

Pseudomonas fluorescens and *A. hydrophila* are the commonly known pathogenic bacteria in freshwater animals and occur naturally in culture environments (Abdel-Aziz, 1988; Roberts, 1989). However, the low incidence of recovery of these bacteria in infected samples (<8%) in contrast to the significant incidences of recovery of *V. harveyi* biotypes (>45.0%) strongly supports the environmental selection and domination of *V. harveyi* in mariculture environments. This finding also emphasizes the significance of the latter biotype as pathogenic to cultured shrimp.

Recovered *Staphylococcus* species from PL15 and pond water samples were Gram-positive cocci, non-motile, catalase positive and oxidase negative, aerobic and facultative anaerobic, non-haemolytic and coagulase negative organisms. *Staphylococcus* species are generally encountered as saprophytic bacteria in freshwater and marine environment (Austin and Austin, 1989). However, its recovery in this study may be related to euryhaline nature of these isolates (Lennette et al. 1985).

Table 9.3: Water quality parameters observed from the shrimp culture ponds

Water Quality Parameter	*Normal Pond*	*Infected Pond*
Temperature (°C)	25 - 32	27 - 35
Salinity (ppt)	34 - 36	36 - 40
pH	7.5 - 8.5	7.2 - 7.5
Dissolved oxygen (ppm)	5.4 - 6.8	4.3 - 5.6
Ammonia-N (ppm)	0.02 - 0.08	0.22 - 0.43
Nitrite-N (ppm)	0 - 0.118	0.132 - 0.624

Biochemical and growth characteristics of recovered Gram-negative bacterial species (Table 10.2) generally agreed with those recorded by Krieg and Holt (1984), Lennette et al., (1985) and the standards of API 20E system (Bio Merieux, France). However, *V.h.*1 and *V.h.*2 were non-luminescent; hence, our result differ from reports of other authors (Liu et al., 1996; Lavilla-Pitogo et al., 1998; Leano et al., 1998). Only Balows et al., (1992) have previously reported the occurrence non-luminescent strain of *V. harveyi*.

Parasitological and mycological examinations of collected samples gave negative results, supporting the association of recovered *V. harveyi* isolates with the recorded mass mortalities among *P. indicus* broodstock and post-larvae (15).

The pathogenesis of *Vibrio* species infections in marine aquatic animals is multifactorial, where a variety of factors related to the host, environment and the pathogen itself may work in consort to define the nature and extent of the triggered course of infection (Snieszko, 1974). In this regard, the differences between the conditions in infected and non-infected ponds, particularly the water temperature, salinity, ammonia and nitrite (Table 9.3) might be related to the outbreaks of the disease in this case. For instance, the increased values of these parameters may result in stressing and immunocompromising of the shrimps (Roberts, 1989) meanwhile enhancing the environmental selection of virulent *V. harveyi* biotypes (Post, 1987). Consequently, *V. harveyi* rapidly invaded the tissues and organs of shrimp. Sunaryanto and Mariam (1986) reported that environmental factors such as temperature, salinity, pH and organic load might be involved in triggering disease outbreaks caused by luminous bacteria. However, Prayitno and Latchford (1995) recorded that exposure of *V. harveyi* to low salinity (10 ppt) significantly increased its virulence while exposure of the bacteria to low pH (5.5) had the reverse effect. Farghaly (1950) demonstrated that environmental factors such as low salinity and fluctuating pH reduced the growth rate of luminous bacteria. Fluctuating environmental factors may induce or enhance the expression of virulence genes in luminous bacteria. Roberts (1989) pointed out that the presence of ammonia is a potential danger to fish health, especially in intensive mariculture systems with pH range of 7.8 to 8.2 and ammonia levels of >0.02 mg/l. However, Williams (1971) recorded that even lower levels of ammonia induced bronchial hyperplasia, while nitrite production by partial oxidation of ammonia can sometimes cause severe losses in fish.

The recovered *V. harveyi* isolates in this study exhibited antibiotic sensitivity to oxytetracycline, nalidixic acid, oxolinic acid, chloramphenicol and enrofloxacin and resistance to sulphonamide, novobiocin, kanamycin, gentamycin and ampicillin. These results suggest the possibility of using antibiotics in disease control. However, there are several problems associated with antibiotic usage including potential environmental and human health hazards (Kerry et al., 1995; Capone et al., 1996) and the possible development and spread of antibiotic-resistant strains (Karunasagar et al. 1994).

REFERENCES

Abdel-Aziz, E.S. 1988. Some studies on Bacterial agents causing tail and fin rot among freshwater fishes in Egypt. M.V.Sc., Thesis. Cairo University.

Austin, B. and Austin, D.A. 1989. Methods for the microbiological examination of fish and shellfish : Ellis Horwood, Chichester.

Austin, B. and Austin, D.A. 1993. Bacterial fish pathogens, 2nd edn: Ellis Horwood Chichester.

Balows, A., Truper, H.G., Dworkin, M., Harder, W. and Schleifer, K.H. 1992. *The Prokaryotes*, Springer-Verlag, New York, Vol. 3 pp. 182.

Baticados, M.C.L., Lavilla-Pitogo, C.R., Cruz-Lacierda, E.R., de la Pena, L.D. and Sunaz, N.A. 1990. Studies on the chemical control of luminous bacteria *Vibrio harveyi* and *Vibrio splendidus* isolated from diseased *Penaeus monodon* larvae and rearing water. Dis. Aquat. Organisms, 9: 133-139.

Bullock, G.L., Conroy, D.A. and Snieszko, S.F. 1971. Bacterial diseases of fishes. In: *Diseases of Fishes*. (eds) S.F. Snieszko and H.R. Axelrod, T.F.H. Publishers, Neptune, pp. 151.

Capone, D.G., Weston, D.P., Miller, V. and Shoemaker, C. 1996. Antibacterial residues in marine sediments and invertebrates following chemotheraphy in aquaculture. Aquaculture 145: 55-75.

Chanratchakool, P., Pearson, M., Limsuwan, C. and Roberts, R.J. 1995. Oxytetracycline sensitivity of *Vibrio* species isolated from diseased black tiger shrimp, *Penaeus monodon* (Fabricius). J. Fish Dis. 18: 79-82.

Chen, S.N. 1992. Coping with diseases in shrimp farming. In: Proc. Global Conf. Shrimp Industry, (eds.) de Saram, H. and Singh, T., Malaysia. Infofish, Kuala Lumpur. pp. 259-272.

Chen, S.N., Huang, S.L., and Kou, G.H. 1992. Studies on the epizootiology and pathogenicity of bacterial infections in cultured giant tiger prawns, *Penaeus monodon*, in Taiwan. In: Diseases of cultured penaeid shrimp in Asia and the United States (eds) Fulks, W. and Main K.I., The Oceanic Institute, Hawaii. pp. 195–205.

Dass, S.M. 1989. Studies on Prawn Cultivation. Ph.D. Thesis, Madurai Kamaraj University. p. 192.

Duijn, C.V.F. 1973. *Diseases of Fishes* liffe Books, London. p. 282.

Farghaly, A.H. 1950. Factor influencing the growth and light production of luminous bacteria. J. Cell. Comp. Physiol. 36: 165-184.

Hjeltnes, B. and Roberts, R.J. 1993. Vibriosis. In: Bacterial Diseases of Fish. (eds) Inglis V, Roberts, R.J. and Bromage, N.R.: Blackwell Scientific Publications, Oxford pp. 109–21.

Jiravanichpaisal, P., Miyazaki, T. and Limsuwan, C. 1994. Histopathology, biochemistry, and pathogenicity of *Vibrio harveyi* infecting black tiger prawn *Penaeus monodon*. J. Aquat. Ani. Hlth. 6: 27-35.

Karunasagar, I., Pai, R., Malathi, G.R. and Karunasagar, I. 1994. Mass mortality of *Penaeus monodon* larvae due to antibiotic-resistant *Vibrio harveyi* infection. Aquaculture 128: 203–209.

Kerry, J., Hiney, M., Coyne, R., NicGabhaim, S., Gilroy, O., Cazabon, D. and Smith, P. 1995. Fish feed as a source of oxytetracycline resistant bacteria in the sediments under fish farms. Aquaculture 131: 101-113.

Koneman, E.W., Allen, S.D., Dowell, V.R. Jr. Janda, W.M. Sommers, H.M. and Winn, W.C. Jr. 1988. Diagnostic microbiology,. JB Lippincott. Philadelphia. pp. 116–148.
Krieg, N.R. and Holt, J.D. (1984). *Bergey's Manual of Systematic Bacteriology*, Williams and Wilkins, Baltimore, Vol.1. pp. 124.
Lavilla-Pitogo, C.R., Baticados, M.C.L., Cruz-Lacierda, E.R. and de la Pena, L.D. 1990. Occurrence of luminous bacterial disease of *Penaeus monodon* larvae in the Philippines. Aquaculture, 91: 1-13.
Lavilla-Pitogo, C.R., Leano, E.M. and Paner, M.G. 1998. Mortalities of pond-cultured juvenile shrimp, *Penaeus monodon*, associated with dominance of luminescent *Vibrios* in the rearing environment. Aquaculture. 164: 337-349.
Leano, E.M., Lavilla-Pitogo, C.R. and Paner, M.G. 1998. Bacterial flora in the hepatopancreas of pond reared *Penaeus monodon* juveniles with luminous vibriosis. Aquaculture 164: 367–374.
Lennette, E.H., Balows, A., Hausler, W.J. Jr. and Shadomy, H.J. 1985. *Manual of Clinical Microbiology*, American Society for Microbiology, Washington, D.C. pp. 154.
Lightner, D.V. 1988. Vibrio disease of penaeid shrimp. In: *Disease diagnosis and control in North American marine aquaculture and fisheries science* (eds) Sinderman, C.J. and Lightner, D.V, Elsevier, Amsterdam. 6: 42-47.
Liu, P.C., Lee, K.K., Yii, K.C., Kou, G.H. and Chen, S.N. 1996. Isolation of *Vibrio harveyi* from diseased Kuruma prawns *Penaeus japonicus*. Curr. Microbiol. 33: 129-132.
Mohaney, L.L., Lightner, D.V. and Bell, T.A. 1994. An epizootic of vibriosis in Ecuadorian pond-reared *Penaeus vannamei* Boone (Crustacea:Decopoda). J. World Aquacult. Soc. 25: 116-125.
Nash, G., Nithimathachoke, C., Tungmandi, C., Arkarjamorn, A., Prathanpipat, P. and Ruamthaveesub, P. 1992. Vibriosis and its control in pond-reared *Penaeus monodon* in Thailand. In: *Diseases in Asian Aquaculture I, Fish Health Section* (eds) Shariff, M., Subasinghe, R.P., and Arthur, J.R, Asian Fisheries Society, Manila, pp. 143-155.
Pizzutto, M. and Hirst, R.G. 1995. Classification of isolates of *Vibrio harveyi* virulent to *Penaeus monodon* larvae by protein profile analysis and M13 DNA fingerprinting. Dis. Aquat. Organisms 21: 61-68.
Post, G. 1987. Textbook of Fish Health. T.F.H. Publications p. 256.
Prayitno, S.B., and Latchford, J.W. 1995. Experimental infections of crustaceans with bacteria related to *Photobacterium* and *Vibrio*. Effect of salinity and pH on infectiosity. Aquaculture, 132: 105–112.
Roberts, R.J. (1989). Fish Pathology. 2nd ed., Bailliere Tindall, London.
Ruangpan, L., Tabkaew, R. and Sangrungruang, K. 1995. Bacterial flora of ponds with different stocking densities of black tiger shrimp, *Penaeus monodon*. In: Diseases in Asian Aquaculture I, Fish Health Section (eds) Shariff, M., Subasinghe, R.P., and Arthur, J.R, Asian Fisheries Society, Manila, pp. 141-149.
Ruby, E.G. and Morin, J.G. 1979. Luminous enteric bacteria of marine fish: A study of their distribution, densities and dispersion. Appl. Environ. Microbiol. 38: 406-411.
Rukyani, A., Taufik, P. and Taukhid, 1992. Penyakit Kunang-Kunang (luminescent vibriosis) di hatchery udang windu dan cara penanggulangannya. Proc. Sem. Penanggulangan Penyakit Udang, Surbaya, Indonesia, pp. 1-17 (in Indonesian).
Snieszko, S.F. 1974. The effects of environmental stress on outbreaks of infectious diseases of fishes. J. Fish. Biol., 6: 197-208.
Sunaryanto, A. and Mariam, A. 1986. Occurrence of a pathogenic bacteria causing luminescence in penaeid larvae in Indonesian hatcheries. Bull. Brackishwater Aquacult. Dev. Centre, 8 : 64-70.
Takahashi, Y., Itami, T., Kondo, M., Maeda, M., Fujii, R., Tomonaga, S., Supamattaya, K., and Boonyaratpalin, S. 1994. Electron microscopic evidence of bacilliform virus infection in Kuruma shrimp (*Penaeus japonicus*). Fish Pathology 29: 121–125.
Williams, W.G. 1971. Nitrite toxicity in a recirculating salmonid culture system. Proc. 22nd Am. N.W. Fish Culture Conf., 80–82.
Yu, S.R. 1995. Studies on the characteristics of *Vibrio alginolyticus* and the lethal factor of the bacteria in *Penaeus monodon* and *P. japonicus*. MS Thesis, National Taiwan Ocean University.

Chapter 10

DNA Vaccine Technology: Current Status and Potential Application in Aquaculture Industry

J.A. Christopher John*, **S. Murali** [1], **M. Peter Marian**[2] and **Chi-Yao Chang**
Molecular Genetics Laboratory, Institute of Zoology
Academia Sinica, NanKang, Taipei 11529, Taiwan
[1]Department of Biological Sciences
National University of Singapore,Singapore
[2]Marine Biotechnology Laboratory, ICAS
S.C. College Campus, Nagercoil, India
Email: john@gate.sinica.edu.tw

ABSTRACT

The possibility of transfecting animal cells in vivo with plasmid DNA encoding antigenic proteins allows the induction of immune responses and the development of a novel method of vaccination. This method, initially called nucleic acid/DNA immunization, has now been employed to elicit protective immunity for viral, bacterial and parasitic diseases in a wide variety of animal models. As infectious diseases have become a major problem in aquaculture, this method may be important for the future development of this industry. This chapter provides a brief account of the current knowledge on the construction of plasmids, methods of administration, safety issues and advantages of DNA vaccines.

Key Words: DNA vaccine, construction of plasmids, mode of administration, safety issues

INTRODUCTION

DNA vaccines are recombinant plasmids that carry a gene for a major antigenic protein of a pathogen. They have received considerable attention as a new approach to vaccine development, especially where traditional vaccines have failed. DNA vaccines offer the advantage of mimicking a pathogenic infection, resulting in the production of a single pathogenic protein that is correctly folded and elicits both cellular and humoral immune responses in the host (Cohen, 1993; Wolff et al., 1990). Generally, DNA vaccines consist of a bacterial plasmid with a strong promoter—usually of viral origin—the gene of interest and a poly-adenylation or transcriptional termination sequence. This plasmid, as amplified in bacteria, is purified, dissolved in a saline solution and then simply injected into the host. The DNA plasmid is taken up by host cells where the encoded protein is created (Donnelly et al. 1997). This expressed protein acts as a vaccine.

Aquaculture, or the farming of aquatic animals, is a rapidly growing industry. The increased demand for fishery products, due to health awareness of consumers and the decline or near stagnation of the natural harvests, have largely contributed to this rapid growth (Meyer, 1991). However, the aquaculture industry needs to augment its global production and efficiency to meet the demand, which is expected to be 10-30% higher than current levels by 2010 AD. A major constraint to the development of this industry is the outbreak of infectious diseases. Although currently available vaccines offer the most efficient way to control infectious pathogens, there are still several important diseases—mainly of viral and parasitic origin—for which no prophylactic treatment exists (Schnick et al., 1997; Heppell and Davis, 2000).

The development of cheap and effective vaccines for the prevention of infectious diseases in finfish has been proved to be a difficult task (Evelyn, 1997; Leong et al., 1997). Recently, several research groups have exhibited the potential of DNA vaccines in aquaculture. Several characteristics of nucleic acid immunization make this technology especially attractive for the prevention of diseases in aquaculture (Leong et al. 1997). Table 10.1 summarizes a brief history of the development of DNA vaccines in aquaculture.

DNA vaccine technology in aquaculture has so far been focused on viral vaccines only. It is most advanced with regard to infectious hematopoietic necrosis virus (IHNV), a major viral disease of salmonids in the USA. An experimental DNA vaccine, encoding the entire glycoprotein of IHNV in a plasmid expression system using a human cytomegalovirus promoter (pCMV-IHNV), induces protection after 5 μg is injected into the muscle of 1 g trout (Anderson et al., 1996b; Leong et al., 1997). DNA vaccines have also been developed for fish viruses, including viral hemorrhagic septicemia virus (VHSV) (Boudinot et al. 1998), spring viremia of carp virus (SVCV) and snakehead rhabdovirus (SHRV) (Kim et al., 2000). Also, these vaccines have been proved to be more effective in protecting fish from lethal challenge with homologous virus than either the traditional killed virus or the subunit vaccine (Leong et al. 1995).

ADVANTAGES OF DNA VACCINES

DNA vaccine technology offers potential solutions to some of the challenges of vaccine development and promises significant advantages over current vaccination strategies. DNA vaccines are relatively simple and inexpensive to prepare. The ease of cloning allows vaccines to be modified rapidly, if needed (Heppell and Davis, 2000). They have been shown to be more stable and resistant to extremes of temperature rather than protein vaccines and, therefore, do not need to be maintained in a cold environment during shipment or storage. Also, they are safer than killed and live vaccines since the risks associated with contaminated vaccines or reversal of live vaccines to virulence are minimized or eliminated (Gómez-Chiarri and Chiaverini, 1999). Moreover, DNA vaccines, have the following advantages. They: (1) generate cytotoxic T-lymphocyte responses that recognize epitopes (antigenic determinants) from conserved proteins and thus provide broader protection against different strains of pathogens; (2) present antigen with native post–translational modifications, conformation and oligomerization to elicit antibodies of optimal specificity; (3) provide long-lasting immune responses; (4) elicit helper T cell responses; and (5) allow the combination of diverse immunogens into a single preparation so as to facilitate simultaneous immunization for several diseases (Donnelly et al., 1997).

CONSTRUCTION OF DNA VACCINES

Vectors

Plasmid constructs used for DNA vaccination purposes, generally consist of the basic attributes of vectors developed for expression of genes in the transfected cell lines. These may include: (1) an origin of replication (*ori* site) for producing high yields of plasmid in *Escherichia coli*; (2) an antibiotic selection marker; (3) a strong enhancer/promoter; (4) an intron; and (5) a polyadenylation signal. Various promoters have been tried and assessed for their ability to drive expression of foreign gene in fish tissues. For example, cytomegalo virus (CMV) immediate early promoter (Anderson et al., 1996a; Heppell et al., 1998; Lorenzen et al., 1998; Boudinot et al., 1998), glucocorticoid-responsive mouse mammary tumor virus (MMTV) promoter (Anderson et al., 1996a), carp β-actin (Anderson et al., 1996a) and herpes simplex virus thymidine kinase (tk) promoter with the CMV enhancer (Gómez-Chiarri et al., 1996) were tested in rainbow trout (*Onchorhynchus mykiss*). SV 40 early promoter was tested in tilapia (*Oreochromis niloticus*) (Rahman and Maclean, 1992); rabbit β-cardiac myosin heavy chain (MHC) and human MxA promoters were tested in common carp (*Cyprinus carpio*) (Hansen et al. 1991). Among these, CMV promoter was found to be the best promoter. It is now widely used for the construction of DNA vaccines reported to date and its potency has been demonstrated in fish vaccination trials (Anderson et al., 1996b; Heppell et al., 1998; Lorenzen et al., 1998, 1999). However, the Food and

Drug Administration does not encourage the use of viral promoters out of safety concerns. In this context, there is an increasing demand for the development of expression vectors constructed with promoter sequences that would be more acceptable for commercial purposes (Hallerman and Kapuscinski, 1995).

Various reporter genes have been used for detecting the expression of injected DNA vaccine. The CMV-promoter-driven lacZ reporter gene (pCMV-LacZ) codes for *E. coli* β-galactosidase (β-gal) is widely used as a model for DNA immunization studies in fish because β-gal is easily visualized in muscle fibres (Davis et al., 1997; Russell et al., 1998). Luciferase gene is also used as a reporter gene, for example, in rainbow trout (*O. mykiss*) (Anderson et al., 1996a), Atlantic salmon (*Salmo salar*) (Gómez-Chiarri and Chiaverini, 1999) and in zebrafish (*Danio rerio*) (Heppel et al., 1998). Different luciferase genes such as firefly (*Photinus pyralis*) luciferase and sea pansy (*Renilla reniformis*) luciferase are available and the extent of gene expression can be assayed by using commercially available kits and a luminometer. Rahman and Maclean (1992) have used chloramphenicol acetyl transferase (CAT) as reporter gene in tilapia (*O. niloticus*). Green and red fluorescence and alkaline phosphatase genes are also being used for the construction of DNA vaccines (unpublished data).

Table 10.1: A brief history of the development of DNA vaccines to fish

Year	*Contributions*
1987	Koener et al. Sequenced glycoprotein cDNA from IHNV. The size was 1609 bp encoding 508 amino acids.
1989	Engelking and Leong, Showed that glycoprotein of IHNV elicits neutralizing Ab and protective responses.
1990	Koener and Leong, Expressed glycoprotein gene by using Baculovirus vectors.
1993	Wolff et al. Transferred gene directly into mouse muscle in vivo.
	Lorenzen et al. Cloned and expressed the glycoprotein gene of VHSV in *E.coli* and immunized rainbow trout with the recombinant protein.
1996	Anderson et al. Developed the FIRST DNA Vaccine for fish against IHNV. This DNA vaccine contained gene coded for the glycoprotein of IHNV.
1998	Boudinot et al. Demonstrated a combined DNA immunization with the glycoprotein gene of VHSV and IHNV.
1999	Corbeil et al. Developed a DNA vaccine with G-gene of IHNV against IHNV.
2000	Corbeil et al. Improved the efficiency of above DNA vaccine.

PATHOGENIC PROTEIN/ANTIGEN-ENCODING GENE

Antigen-encoding gene is the major element required for the construction of a DNA vaccine and also for stimulating specific immunity against a particular pathogen. Theoretically, any gene coding for a protein of a pathogen that induces a protective immune response in the host can be used in DNA vaccines, as long as the codon usage of the gene allows expression in fish cells. Sometimes, a problem may eventually arise with the codon usage of the genes from some bacteria or parasites, but it is unlikely to happen with viral genes, as they normally use the host machinery for expression. DNA vaccines may also include a signal peptide to direct secretion of the encoded antigen, although this does not appear to be an absolute requirement (Heppell and Davis, 2000).

DNA AS AN IMMUNOSTIMULANT

Besides its function of encoding the genetic material, DNA can have direct immune stimulatory effects. It is not surprising that the polyanionic polymer structure of DNA can interact with other biological molecules in complex ways (Krieg, 2001). The specific immunostimulatory effect of bacterial genomic DNA was first reported by Tokunaga and his colleagness who demonstrated that bacterial DNA activates natural killer (NK) cells and induces interferon (IFN) production, but vertebrate DNA does not (Tokunaga et al., 1984; Yamamoto et al. 1988, 1992). Also, it has been proved that bacterial DNA, but not vertebrate DNA, activates B cell proliferation and immunoglobulin secretion (Messina et al. 1991).

The ability of an antibody to neutralize viruses is associated with its avidity (Blank et al. 1972). Antibody avidity is 100-1000-fold higher in mice when the antibody is induced by intramuscular DNA as compared to protein in saline, although high avidity antibody is also induced by protein bound to alum adjuvant (Boyle et al. 1997). It has been shown that 50 μg intramuscular injection of plasmid DNA in saline and intraperitoneal injection of 10μg β-gal protein in a commercial oil adjuvant induced the same serum antibody titre and avidity in gold fish (Kanellos et al. 1999a).

CpG motifs

An unmethylated, six-nucleotide sequence consisting of two purine residues followed by CG and two pyrimidine residues (RRCGYY) called CpG motifs, has been reported to play an important role in the stimulation of potent immune response (Lai et al. 1995). These oligonucleotides induce secretion of interleukins (ILs) as IL-1, IL-2, IFN-γ, and tumor necrosis factor alpha. Bacterial DNA differs from vertebrate DNA in having a much higher content of unmethylated CpG dinucleotides. The vertebrate immune system recognizes 'CpG motifs' as foreign and triggers protective immune responses, which have thus shown remarkable utility as vaccine adjuvants (Krieg, 2001).

These CpG motifs can be deliberately added to DNA vaccines in order to enhance the Th1 immune response. B cell activation can also be triggered by synthetic oligodeoxynucleotides (ODNs), which contain an unmethylated CpG dinucleotide in a particular sequence context. This B cell activation is T-cell independent and antigen non-specific (Krieg et al. 1998). It has also been shown that CpG DNA activates monocytes, macrophages and dendric cells in vitro to upregulate their expression of co-stimulatory molecules that drive immune responses and to secrete a variety of cytokines, including high levels of interleukin 12 (IL-12) (Klinman et al., 1996; Halpern et al., 1996; Chace et al., 1997). These cytokines stimulate natural killer cells to secrete IFN-γ and have increased lytic activity (Yamamoto et al., 1992; Klinman et al., 1996, Cowdery et al., 1996).

Several studies have shown the ability of CpG DNA to act as an adjuvant for infectious disease vaccines such as hepatitis B vaccine in mice (Davis et al., 1998) and orangutans (Davis et al. 2000). Although fish DNA vaccines contain CpG motifs (example, Kim et al., 2000), their influence in the immune response has not been reported.

G gene

The glycoprotein G of the fish virus IHNV is a membrane-associated protein which forms spike-like projections on the surface of the mature virion (McAllister and Wagner, 1975). It is responsible for the attachment to the host cell membrane and the entry within the cell by a receptor-mediated endocytosis (Lecocq-Xhonneux et al. 1994). Koener et al., (1987) first sequenced the glycoprotein cDNA from IHNV in 1987. Ever since, glycoproteins have been used in vaccine development for fish. The first DNA vaccine for fish against IHNV (Anderson et al., 1996b) as well as all other vaccines so far reported contains G-gene (Table 10.1).

Although G-gene containing DNA vaccines are found to be successful, the mechanism of protection remains unclear. Recently, Kim et al., (2000) reported that DNA vaccination with a plasmid encoding a viral G resulted in the production of IFN-α, which provided an early and effective protection against virus infection. Over a period of time, the non-specific anti-viral state was replaced by the long-term specific protection provided by the DNA vaccine through a cellular immune response (Kim et al. 2000).

It is also interesting to note that the glycoprotein of one isolate of IHNV can stimulate protective immunity in fish against various strains of the virus. Hence, the G gene containing DNA vaccines might have widespread utilization.

METHODS OF ADMINISTRATION

DNA vaccination to aquatic animals can be administered by one of three ways: intramuscular or intraperitoneal injection, immersion of the fish in a vaccine solution and oral administration. These methods have different advantages with respect to the level of protection, side-effects, practicality and cost efficiency.

Injection

Site of injection: The results of several studies revealed that injection is the most effective method to immunize fish (for example, Newman, 1993; Press and Lillehaug, 1995). But this method is laborious and stressful for the animals. Purified plasmid DNA in a small volume of saline or buffer is usually injected intramuscularly in the flank, below or close to the dorsal fin, gills or the eye (Table 10.2). Vaccination via intraperitoneal injection induced partial protection in rainbow trout against IHNV (Corbeil et al. 2000b). Kanellos et al., (1999a) have reported that DNA immunization of goldfish *via* intraperitoneal injection induces much higher antibody titres when the vaccine solution contains oil adjuvants.

Table 10.2: Methods of injection, sites and doses of DNA vaccines tested in various species of fish (Modified from Heppell and Davis, 2000)

Method[a]	*Fish*	*Site*	*Gene*[c]	*Dose (μg)*	*Volume (μl)*	*Reference*
IM single injection	Rainbow trout	Flank[b]	pCMV4-G pCMV4-N	10 10		Anderson et al. (1996b) Anderson et al. (1996b)
	Rainbow trout	Flank	VHSV-G&N	5,10 and 50	25	Lorenzen et al. (1998)
			VHSV-G&N			
	Rainbow trout	Flank		0.1,1 and 10	25	Lorenzen et al. (1999)
			IHNVw-G			
	Rainbow trout	Dorsal fin	IHNVw-G	100 ng	50	Corbeil et al. (2000b)
	Rainbow trout	Dorsal fin	VHSV-G IHNV-G and	1-10 ng	50	Corbeil et al. (2000a)
IM multiple injection	Rainbow trout	Flank	rabies virus	30	100	Boudinot et al. (1998)
			IHNVw-G			
IP	Rainbow trout	Ventral fin	IHNVw-G	100 ng	50	Corbeil et al. (2000b)
Particle bombardment	Rainbow trout	Dorsal fin		100 ng	NA[d]	Corbeil et al. (2000b)

[a]IM - intramuscular; IP – intraperitoneum
[b]Flank: The exact location of injection site varies, usually located below or rostroventral to the dorsal fin and above the lateral line.
[c]Genes: CMV – cytomegalovirus; VHSV – viral haemorrhagic septicaemia virus and IHNVw – infectious hematopoietic necrosis virus WRAC strain (Corbeil et al., 1999; 2000 a, b)
[d]Not applicable

Dose and volume: A number of reports are available on the effect of dose and volume of the injected DNA, as they play an important role in the expression of gene in the DNA vaccine. It has been shown that higher doses and larger volumes may reduce the expression level of foreign gene and reporter gene activity (Gómez-Chiarri et al., 1996; Schulte et al., 1998). Optimum doses fall in the range of 1-50 µg of DNA in the volume of 10-50 µl. However, much lower doses have also been found to be effective (Heppell et al. 1998). In fact, doses as low as 0.1 µg of DNA per fish have been reported to be as effective as 10 µg for inducing protective immunity (Lorenzen et al. 1999). Recently, Corbeil et al. (2000a) showed that 1-10 ng of DNA was sufficient to induce remarkable immunity. The optimal dose of DNA may vary according to the species and age of the fish. Hansen et al. (1991) revealed that young and fast growing carp showed much higher levels of chloramphenicol acetyl transferase (CAT) reporter gene activity than older fish.

Location and duration of expression of injected gene

In large fish, expression of reporter genes seems to be mostly restricted to the site of injection, i.e. in myocytes as well as in cells infiltrating the muscle tissue and epithelial cells lining small capillaries (Boudinot et al. 1998). In contrast, expression of reporter genes in small fish was detected in other organs such as the gills, spleen and kidney as well as in the site of injection (Anderson et al., 1996a; Heppell et al., 1998). Differences between large and small fish are most likely due to more rapid spreading of injected DNA from the site of injection in small fish, but there may have been also variations in the protocol, sensitivity of the assay and animal model used (Heppell and Davis, 2000).

PARTICLE BOMBARDMENT

Particle bombardment has been tried in fish as an alternative method for DNA injection (Gómez-Chiarri et al., 1996; Corbeil et al., 2000b). By this method, the DNA vaccine is administered into the dermis and muscle of young fish (Gómez-Chiarri et al. 1996). However, this method was not as efficient as needle injection both in terms of the amount of an individual's expression and the proportion of animals expressing the reporter gene. Also, this technology cannot be widely employed on fish farms because of its high cost and practical limitations (Heppell and Davis, 2000).

IMMERSION AND ORAL ADMINISTRATION

Oral administration of vaccines is the ideal method of immunization, as it is easy, inexpensive to administer and also not stressful for the animals. Induction of mucosal immunity is an immunological advantage associated with oral delivery (Zapata et al. 1996). There is still a limited understanding of the mechanisms involved in antigen uptake and presentation after oral vaccination (Nakanishi and Ototake, 1997a; Quental and Vigneulle, 1997; Moore et al., 1998). So far, research has been focused on protecting the antigens from digestion and decomposition during passage through the stomach and anterior part of the gut (Gudding et al. 1999). This immunization method can be widely used in large fish farms. By this method, it is not possible to determine the dose of vaccine ingested by each animal and so it appears to be less effective than injection or immersion for vaccinating fish.

The studies of Corbeil et al., (2000b) showed that rainbow trout fry immunized with the pIHNVw-G DNA vaccine through the mouth were not protected against viral challenge. The unprotected DNA may have been damaged by the digestive tract and/or not taken up by the cells. Chen et al. (1998) achieved induction of immunity in mice by oral immunization with encapsulated DNA. In this study, the DNA was incorporated into biodegradable polyactide-coglycolide microspheres that protected the vaccine as it passed through the digestive tract. The microspheres may also have facilitated the DNA uptake by the immune cells. A similar strategy may also be successful in fish and is worth investigating in future studies. One possibility is the incorporation of DNA vaccine-microencapsules in food items such as brine shrimp (*Artemia).*

Immersion vaccination involves any one of the following delivery techniques such as direct immersion, hyperosmotic dip, flush exposure and spraying of animal with a concentrated vaccine solution. Besides phagocytes, several types of epithelial cells are involved in antigen uptake (Nakanishi and Ototake, 1997b). But no information is available to date on administration of DNA vaccines to fish by immersion methods. Heppell et al., based on their unpublished data, suggest that immersion or oral vaccination with formulated DNA vaccines might be feasible and effective, when the vaccines are transferred through liposomes or microcarriers to fish (Heppell and Davis, 2000).

ELECTROPORATION

Genetronics research paves the way to use electrically assisted methods such as in vivo electroporation for gene delivery. Due to its accessibility and regenerative capabilities, this in vivo gene delivery system becomes attractive for DNA vaccination. Recently, Genetronics studies have been aimed at: (1) Examining the location of transgene expression in skin; (2) Determining the level of transgene expression and secretion of transgene products as a result of injecting naked plasmid DNA into skin, followed by non-invasive in vivo electroporation; (3) Comparing the characteristics of gene expression in skin and muscle; and (4) Evaluating the immune response elicited by transgene expression after delivering the DNA into the skin and muscle *via* electroporation. It was also demonstrated that in vivo electroporation technology is a simple and extremely effective method for delivering naked DNA into both skin and muscle (Zhang, 2000). This method may also be tried in fish for effective DNA vaccination.

SAFETY CONSIDERATIONS

The approval of DNA vaccines for commercial use in aquaculture poses several regulatory challenges (Sethi et al. 1997). Hence, special consideration must be given to safety issues and the potential economical benefits must be evaluated with respect to potential negative effects on the host (fish), the consumer (human) and the environment (Gómez-Chiarri and Chiaverini, 1999).

FISH

The considerations about safety of DNA vaccines are different for food source animals and humans (Yamanouchi et al. 1998). In food source animals, there is less concern about integration of DNA into the genome and the possibility of induction of autoimmune diseases. Integration of plasmid DNA into the host genome is very unlikely as it is the risk of an insertional mutagenic event. No such incidents have been detected so far, even with sensitive PCR methods (Nichols et al., 1995; Anderson et al., 1996a; Kanellos et al., 1999b).

The DNA administered to fish is very stable. Studies revealed that plasmids injected in muscle could be detected even at 63 days in rainbow trout (Anderson et al., 1996a) and 70 days in gold fish (Boudinot et al., 1998; Kanellos et al., 1999b) after injection. The persistence of plasmid DNA is associated with long-term expression of the encoded antigen which induces antibody and memory T-like cells in a safe manner (Kanellos et al. 1999b).

CONSUMER

The only possible risk to the consumer may be associated with the absorption of plasmid DNA from the injected fish. However, most of this plasmid DNA would be degraded by the time the flesh is cooked. Even if the fish were eaten immediately after injection, the amount of plasmid DNA consumed would be negligible compared to the total amount of nucleic acid from all other sources—including bacteria and viruses—normally ingested daily. Moreover, common vaccines made up of inactivated or attenuated pathogens also contain large amounts of nucleic acids, mostly from the pathogens themselves. Hence, it could be concluded that, DNA vaccines are no more dangerous than the currently used vaccines (Heppell and Davis, 2000).

ENVIRONMENT

The injected plasmids that fail to enter the nucleus of cells are rapidly degraded by nucleases in the extracellular space and the cytoplasm (Heppell and Davis,

2000). However, a small portion of DNA may remain for prolonged periods in the site of injection. It was initially thought that these free foreign genes could cause the host to become tolerant to the gene product. This would affect not only the immunized fish, but also other fish, since tolerant animals could become carriers of the infectious agents. Nevertheless, tolerance has never been demonstrated in DNA-immunized animals so far (Davis and McCluskie, 1999).

DNA vaccinated fish are not transgenic animals. Also, it is unlikely that the injected plasmid could pass on to other organisms or to the next generation of fish. Moreover, as the DNA sequence encodes only a single viral gene, there should be no possibility of reversion to virulence, which is a critical factor in relation to environmental safety in aquaculture (Gudding et al. 1999).

CONCLUSION

Research on the stimulation of non-specific and specific immunity in fish may be fundamental for developing aquaculture industry into a sustainable source of high quality animal protein. The development of safe and efficient expression vectors for the construction of DNA vaccines against infectious diseases in aquaculture is an immediate requirement. Hence, future research must evaluate thoroughly the activity of different promoters in fish species of commercial interest.

REFERENCES

Anderson, E.D., Mourich, D.V., and Leong, J.C. 1996a. Gene expression in rainbow trout (*Oncorhynchus mykiss*) following intramuscular injection of DNA. Mol. Mar. Biol. Biotechnol. 5: 105-113.

Anderson, E.D., Mourich, D.V., Fahrenkrug, S.C., LaPatra, S., Shepherd J., and Leong. J.C. 1996b. Genetic immunization of rainbow trout (*Oncorhynchus mykiss*) against infectious hematopoietic necrosis virus. Mol. Mar. Biol. Biotechnol. 5: 114-122.

Blank, S.E., Leslie, G.E. and Clem. L.W. 1972. Antibody affinity and valence in viral neutralization. J. Immunol. 108: 665-673.

Boudinot, P., Blanco, M. de Kinkelin P. and Benmansour, A. 1998. Combined DNA immunization with the glycoprotein gene of viral hemorrhagic septicemia virus and infectious hematopoietic necrosis virus induces double-specific protective immunity and nonspecific response in rainbow trout. Virology, 249: 297-306.

Boyle, J.S., Silva, A., Brady J.L. and Lew. A.M. 1997. DNA immunization: Induction of higher avidity antibody and effect of route on T cell cytotoxicity. Proc. Natl. Acad. Sci. USA, 94: 14626-14631.

Chace, J.H., Hooker, N.A., Mildenstein, K.L., Krieg A.M. and Cowdery. J.S. 1997. Bacterial DNA induced NK cell IFN- γ production is dependent on macrophage secretion of IL-12. Clin. Immunol. Immunopathol. 84: 185-193.

Chen, S.C., Jones, D.H., Fynan, E.F., Farrar, G.H., Clegg, J.C., Greenberg, H.B. and J.E. Herrmann. 1998. Protective immunity induced by oral immunization with a rotavirus DNA vaccine encapsulated in microparticles. J. Virol. 72: 5757-5761.

Cohen, J. 1993. Naked DNA points way to vaccines. Science, 259: 1691-1692.

Corbeil, S., LaPatra, S.E., Anderson, E.D. Jones, J. Vincent, B., Hsu Y.L. and G. Kurath. 1999. Evaluation of the protective immunogenicity of the N,P,M,NV and G proteins of infectious hematopoietic necrosis virus in rainbow trout *Oncorhynchus mykiss* using DNA vaccines.

Dis. Aquatic Organ. 39: 29-36.
Corbeil, S., LaPatra, S.E., Anderson E.D. and G. Kurath. 2000a. Nanogram quantities of a DNA vaccine protect rainbow trout fry against heterologous strains of infectious hematopoietic necrosis virus. Vaccine, 18: 2817-2824.
Corbeil, S., Kurath G. and LaPatra. S.E. 2000b. Fish DNA vaccine against infectious hematopoietic necrosis virus: Efficacy of various routes of immunization. Fish Shellfish Immunol. 10: 711-723.
Cowdery, J.S., Chace, J.H., Yi A.-K. and Krieg. A.M. 1996. Bacterial DNA induces NK cells to produce interferon-ã in vivo and increases the toxicity of lipopolysaccharides. J. Immunol. 156: 4570-4575.
Davis, H.L. and McCluskie. M.J. 1999. DNA vaccines for viral diseases. Microbes and Infections, 1: 7-21.
Davis, H.L., C.L.B. Millan and S.C. Watkins. 1997. Immune-mediated destruction of transfected muscle fibers after direct gene transfer with antigen-expressing plasmid DNA. Gene Therapy, 4: 181-188.
Davis, H.L., Weeratna, R., Waldschmidt, T.J. Tygrett, L., Schorr J. and Krieg. A.M. 1998. CpG DNA is a potent adjuvant in mice immunized with recombinant hepatitis B surface antigen. J. Immunol. 160: 870-876.
Davis, H.L., Suparto, I., Weeratna, R., Risini, J., Iskandriati, D., Chamzah, S., Ma'ruf, A., Nente, C., Pawitri, D., Krieg, A.M., Heriyanto, Smits W. and Sajuthi, D. 2000. CpG DNA overcomes hypo-responsiveness to hepatitis B vaccine in Orangutans. Vaccine, 18:1920-1924.
Donnelly, J.J., Ulmer, J.B. Shiver J.W. and Liu. M.A. 1997. DNA vaccines. Annu. Rev. Immunol. 15: 617-648.
Engelking, H.M. and Leong. J.C. 1989. The glycoprotein of infectious hematopoietic necrosis virus elicits neutralizing antibody and protective responses. Virus Res.13: 213-230.
Evelyn, T.P. 1997. A historical review of fish vaccinology. Dev. Biol. Stand. 90: 3-12.
Food and Drug Administration. 1996. Points to consider on plasmid DNA vaccines for preventive infectious disease indications. http://www.fda.gov/cber/ptc/plasmid.txt
G6mez-Chiarri, M. and Chiaverini. L.A. 1999. Evaluation of eukaryotic promoters for the construction of DNA vaccines for aquaculture. Genetic analysis: Biomolecular Engineering, 15: 121-124.
G6mez-Chiarri, M., Livingston, S.K., Muro-Cacho, C., Sanders S. and Levine. R.P. 1996. Introduction of foreign genes into the tissue of live fish by direct injection and particle bombardment. Dis. Aquat. Org. 27: 5-12.
Gudding, R., Lillehaug A. and Evensen. Ø. 1999. Recent developments in fish vaccinology. Veter. Immunol. Immunopathol. 72: 203-212.
Hallerman, E.M. and Kapuscinski, A.R. 1995. Incorporating risk assessment and risk management into public policies on genetically modified finfish and shellfish. Aquaculture, 137: 9-17.
Halpern, M.D., Kurlander R.J. and Pisetsky, D.S. 1996. Bacterial DNA induces murine interferon-γ production by stimulation of interleukin-12 and tumor necrosis factor-α. Cell. Immunol. 167: 72-78.
Hansen, E., Fernandes, K., Goldspink, G., Butterworth, P., Umeda P.K. and Chang, K.-C. 1991. Strong expression of foreign genes following direct injection into fish muscle. FEBS Lett. 290: 73-76.
Heppell, J. and Davis. H.L. 2000. Application of DNA vaccine technology to aquaculture. Adv. Drug Del. Rev. 43: 29-43.
Heppell, J., Lorenzen, N., Armstrong, N.K., Wu, T., Lorenzen, E., Einer-Jensen, K., Schorr J. and Davis. H.L. 1998. Development of DNA vaccines for fish: Vector design, intramuscular injection and antigen expression using viral haemorrhagic septicaemia virus genes as model. Fish Shellfish Immunol. 8: 271-286.
Kanellos, T., Sylvester, I.D., Howard C.R. and Russell, P.H. 1999a. DNA is as effective as protein at inducing antibody in fish. Vaccine, 17: 965-972.
Kanellos, T., Sylvester, I.D., Ambali, A.G., Howard C.R. and Russell, P.H. 1999b. The safety and longevity of DNA vaccines for fish. Immunology, 96: 307-313.
Kim, H.C., Johnson, M.C., Drennan, J.D., Simon, B.E. thomann and Leong. J.C. 2000. DNA vaccines

encoding viral glycoprotein induce non-specific immunity and Mx protein synthesis in fish. J. Virol. 74: 7048-7054.

Klinman, D.M., Yi, A.K., Beaucage, S.L., Conover J. and Krieg, A.M. 1996. CpG motifs present in bacteria DNA rapidly induce lymphocytes to secrete interleukin 6, interleukin 12, and interferon gamma. Proc. Natl. Acad. Sci. USA. 93: 2879-2883.

Koener, J.F. and Leong, J.C. 1990. Expression of the glycoprotein gene from a fish rhabdovirus by using baculovirus vectors. J. Virol. 64: 428-430.

Koener, J.F., Passavant, C.W., Kurath G. and Leong, J.C. 1987. Nucleotide sequence of a cDNA clone carrying the glycoprotein gene of infectious hematopoietic necrosis virus, a fish rhabdovirus. J. Virol. 61: 1342-1349.

Krieg, A.M. 2001. Immune effects and mechanisms of action of CpG motifs. Vaccine, 19: 618-622.

Krieg, A.M., Wu, T., Weeratna, R., Efler, S.M., LoveHoman, L., Yang, L., Yi, A.-E., Short D. and Davis, H.L. 1998. Sequence motifs in adenoviral DNA block immune activation by stimulatory CpG motifs. Proc. Natl. Acad. Sci. USA. 95: 12631-12636.

Lai, W.C., Bennett, M., Johnston, S.A., Barry M.A. and Pakes. S.P. 1995. Protection against *Mycoplasma pulmonis* infection by genetic vaccination. DNA Cell Biol. 14: 643-651.

Lecocq-Zhonneux, F., Thiry, M., Dheur, I., Rossius, M., Vanderheijden, N., Martial J. and Kinkelin, P de 1994. A recombinant viral haemorrhagic septicaemia virus glycoprotein expressed in insect cells induces protective immunity in rainbow trout. J. Gen. Virol. 75: 1579-1587.

Leong, J.C., Bootland, L.M., Anderson, E., Chiou, P.W., Drolet, B., Kim, C., Lorz, H., Mourich, D., Ormonde, P., Perez L. and Trobridge, G. 1995. Viral vaccines in aquaculture. J. Mar. Biotechnol. 3: 16-21.

Leong, J.C., Anderson, E., Bootland, L.M., Chiou, P.W., Johnson, M., Kim, C., Mourich, D., Ormonde, P., Perez L. and Trobridge, G. 1997. Fish vaccine antigens produced or delivered by recombinant DNA technologies. Dev. Biol. Stand. 90: 267-277.

Lorenzen, N., Olesen, N.J., Jorgensen, P.E., Etzerodt, M., Holtet T.L. and Thogersen. H.C., 1993. Molecular cloning and expression in *Escherichia coli* of the glycoprotein gene of VHS virus and immunization of rainbow trout with the recombinant protein. J. Gen. Virol. 74: 623-630.

Lorenzen, N., Lorenzen, E., Einer-Jensen, K., Heppell, J., Wu T. and Davis, H.L. 1998. Protective immunity to VHS in rainbow trout (*Oncorhynchus mykiss*, Walbaum) following DNA vaccination. Fish Shellfish Immunol. 8: 261-270.

Lorenzen, N., Lorenzen, E., Einer-Jensen, K., Heppell J. and Davis, H.L. 1999. Genetic vaccination of rainbow trout against viral haemorrhagic septicaemia virus: Small amounts of DNA protect against a heterologous serotype. Virus Res. 63:19-25.

McAllister, P.E. and Wagner, R.R. 1975. Structural proteins of two salmonid rhabdoviruses. J. Virol. 15: 733-738.

Messina, J.P., Gilkeson G.S. and Pisetsky, D.S. 1991. Stimulation of in vitro murine lymphocyte proliferation by bacterial DNA. J. Immunol. 147: 1759-1764.

Meyer, F.P. 1991. Aquaculture disease and health management. J. Anim. Sci. 69:4201-4208.

Moore, J.D., Ototake M. and Nakanishi, T. 1998. Particulate antigen uptake during immersion immunization of fish: The effectiveness of prolonged exposure and the roles of skin and gill. Fish Shellfish Immunol. 8: 393-407.

Nakanishi, T. and Ototake, M. 1997a. In: Fish Vaccinology. (Eds.) R. Gudding, A. Lillehaug, P.J. Midtlyng and F. Brown. Dev. Biol. Stand. Karger, Basel, p. 59.

Nakanishi, T. and Ototake, M. 1997b. Antigen uptake and immune responses after immersion vaccination. Dev. Biol. Stand. 90: 59-68.

Newman, S.G. 1993. Bacterial vaccines for fish. Ann. Rev. Fish Dis. 3:145-185.

Nichols, W.W., Ledwith, B.J., Manam, S.V. and Troilo, P.J. 1995. Potential DNA vaccine integration into host cell genome. Ann. N.Y. Acad. Sci. 772: 30-39.

Press, C.McL. and Lillehaug, A. 1995. Vaccination in European salmonid aquaculture: a review of practices and prospects. Br. Vet. J. 151: 45-69.

Quentel, C. and Vigneulle, M. 1997. In: Fish Vaccinology. (eds) Gudding, R., Lillehaug, A., Midtlyng P.J. and Brown. F. Dev. Biol. Stand. Karger, Basel, p. 69.

Rahman, A. and Maclean, N. 1992. Fish transgenic expression by direct injection into fish muscle. Mol. Mar. Biol. Biotechnol. 1: 286-289.

Russel, P.H., Kennellos, T., Sylvester, I.D., Chang K.C. and Howard, C.R. 1998. Nucleic acid immunization with a reporter gene results in antibody production in gold fish *(Carassius auratus* L.). Fish Shellfish Immunol. 8: 121-128.

Schnick, R.A., Alderman, D.J., Armstrong, R., Gouvello, R., Le Ishihara, S., Lacierda, E.C., Percival S. and Roth, M. 1997. Worldwide aquaculture drug and vaccine registration progress. Bull. Eur. Ass. Fish Pathol. 17: 251-260.

Schulte,P.M., Powers D.A. and Schartl, M. 1998. Efficient gene transfer into *Xiphophorus* muscle and melanoma by injection of supercoiled plasmid DNA. Mol. Mar. Biol. Biotechnol. 1: 241-247.

Sethi, M.S., Gifford G.A. and Samagh, B.S. 1997. Canadian regulatory requirements for recombinant fish vaccines. Dev. Biol. Stand. 90: 347-353.

Tokunaga, T., Yamamoto H. and Shimada, S. 1984. Antitumor activity of deoxyribonucleic acid fraction from mycobacterium bovis GCG. Isolation, physico-chemical characterization, and antitumor activity. JNCI, 72: 955-962.

Wolff, J.A., Malone, R.W., Williams, P., Chong, W., Ascadi, G., Jani A. and Felgner, P.L. 1990. Direct gene transfer into mouse muscle in vivo. Science, 247: 1465.

Xiang, Z.Q., Spitalnik, S., Tran, M., Wunner, W.H., Cheng J. and Ertl. H.C.J. 1994. Vaccination with a plasmid vector carrying the rabies virus glycoprotein gene induces protective immunity against rabies virus. Virology, 199: 132-140.

Yamamoto, S., Kuramoto, E., Shimada S. and Tokunaga, T. 1988. In vitro augmentation of natural killer cell activity and production of interferon-α/β and -γ with deoxyribonucleic acid fraction from mycobacterium bovis BCG. Jpn. J. Cancer Res. 79: 866-873.

Yamamoto, S., Yamamoto, T., Shimada, S., Kuramoto, E., Yano, O., Kataoka T. and Tokunaga, T. 1992. DNA from bacteria, but not from vertebrates, induces interferons, activates natural killer cells and inhibits tumor growth. Microbiol. Immunol. 36: 983-997.

Yamanouchi, K., Barrett T. and Kai, C. 1998. New approaches to the development of virus vaccines for veterinary use. Rev. Sci. Tech. Off. Int. Epiz. 17: 641-653.

Zapata, A.G., Chiba A. and Varas, A. 1996. Cells and tissues of the immune system of fish. In: The fish immune system: Organism, Pathogen and Environment, (eds) Iwama. G and Nakanishi, T. Academic press, San Diego, pp. 1-62.

Zhang, L. 2000. Electroporation-enhanced delivery of plasmid DNA for gene therapy and DNA vaccines. Abstract of International symposium on DNA vaccine and gene therapy technology, Academia Sinica, Taiwan.

Chapter 11

Isolation of Antibody-like Substances from Marine Algae

U. Barros, and **A. Himanshu**

Department of Biotechnology
Goa University, Goa 403 206, India
email: urmila_goa@sancharnet.in

ABSTRACT

Ulva fasciata, Gracilaria corticata and *Enteromorpha intestinalis* were collected from the shores of Kakra, Anjuna and Cabo de Rama along the Goan coast (Southwestern India) and screened for antibody-like substances. Protein was extracted by ammonium sulphate precipitation and the highest and lowest protien yields were obtained from *U. fasciata* and *G. corticata*, respectively. Each of the protein extracts showed hemagglutination activity against human red blood cells. The extract from *E. intestinalis* showed partial blood group specificity. Hemmagglutination activity of *U. fasciata* extract was inhibited with all 6 sugars tested for competitive inhibition. The crude extract contains a "lectin-like substance".

Keywords: agglutinin, lectin, protein, marine algae, hemagglutination

INTRODUCTION

The presence of proteins with the remarkable ability to agglutinate erythrocytes in plant extract is known to man since the turn of the nineteenth century (Wright and Douglas, 1903). However, only sporadic studies of these agglutinins, called "lectins", were conducted. Recently, various lectins have been isolated and their physiological functions are being understood (Colucci, et al. 1999).

Lectins are proteins with sugar-binding subunits that recognize and bind to specific carbohydrates on the surface of cells. This carbohydrate specificity has rendered them valuable for a variety of applications, especially in the field of immunology. This has stimulated surveys of lectins in higher plants (Etzler, 1986).

In comparison to terrestrial plants, there are few reports of lectins identified from marine plants. The first report of lectins in marine algae was in 1966 by Boyd et al. (1966). Subsequently, sea weeds from Britain (Blunden et al., 1975), Japan (Hori et al., 1987) and America (Bird et al., 1993) have been screened for lectin production and function.

The purpose of this study was to screen some algal species from the coast of Goa (Southwestern India) for the presence of lectins and to determine, if the algal extract (partially purified) produced agglutination of human erythrocytes.

MATERIAL AND METHODS

Sample collection

Three kinds of marine algae were collected from the rocky shores of Anjuna, Kakra and Cabo-de-Rama and further identified as *Ulva fasciata, Gracilaria corticata* and *Enteromorpha intestinalis*. Upon collection, the algae were cleaned of epiphytes, rinsed in tap water and frozen at –20°C. The samples were then lyophilized and stored at –20°C until protein extraction.

Protein extraction

The method used for protein extraction was a modification of the same used by Dalton et al. (1995). All procedures were conducted at 4°C. Approximately 30 g of each sample was homogenized in minimum volume of TBS buffer. The final volume was then made upto 300 ml. The algal homogenate was suspended in 20 volumes (v/v) of TBS and stirred for 18 hrs. The homogenate was first filtered through a muslin cloth and then by vacuum filtration through an ordinary filter paper. The filtrate obtained was saturated to 20% with ammonium sulphate and mixed for 4 hrs using a magnetic stirrer. The supernatant obtained after centrifuging (1000 g for 20 min.) was made upto 75% saturation and stirred for 16 hours. On re-centrifuging (1000g, 20 min.), the final pellet obtained was re-suspended in buffer and dialyzed against 1*l* of 0.01M PBS buffer for 48 hrs. The protein concentration of the final extract was assayed by Folin-Lowry assay and determined from the standard BSA curve.

Hemagglutination assay

Hemagglutination assay was conducted on human Rh+ A, B, AB and O erythrocytes. A 3% RBC suspension (v/v) was made in cold PBS after centrifuging (2000 rpm, 10 min.) and washing thrice in cold PBS (0.1M). Aliquots of 50μl of the algal extracts were thoroughly mixed with equal volumes of 3% erythrocyte suspension in flat-bottomed wells of a micro-titer plate and kept at room temperature for 1 hr. Antisera (anti-A, B and AB) served as positive controls. Agglutination activity was observed after 1 hr. To determine the hemagglutination titer of each extract, the same procedure as above was followed except that the algal extracts were serially diluted twofold until 1/16 dilution. The activity in hemagglutination assays is reported as the minimum amount of protein tested that produces agglutination.

Hemagglutination inhibition assay

Only the extract from *U. fasciata* and Rh+ AB erythrocytes were used to test sugar-induced hemagglutination inhibition. The sugars used were D-glucose, D-galactose, D-fructose, D-xylose, D-mannose and L-rhamnose. To determine the optimum concentration of carbohydrate for inhibition, 50μl of algal extract was incubated with 50μl sugar solutions ranging from 75 mM to 1000 mM, for 15 min. at room temperature and then processed in the same way as for the routine hemagglutination assay described earlier. Mild inhibition was observed with 250 mM and thus, 500 mM was used in further inhibition tests.

RESULTS

The extracts obtained from the 3 algae were confirmed to be protein by Folin-Lowry assay. The highest protein yield (112.7μg/g) was obtained from *U. fasciata* extract, whereas the least yield was from *G. corticata* (72.6μg/g). Results of the hemagglutination assays are reported in Table 11.1.

The titers of the extracts are summarized in Table 11.2. All three protein extracts produced some degree of agglutination in human Rh+ A and AB erythrocytes. Extracts from the protein extract from *U. fasciata* was tested for sugar-binding specificity against AB erythrocytes. Hemagglutination inhibition was observed with all six sugar solutions used in this study.

Table 11.1: Observations of hemagglutination assay
- No agglutination; +/- Partial or not clear;
+ Agglutination (+ weak, ++ moderate, +++ intense)

Blood types Rh+	*Dilution Factor*					*Protein Extract*
	1	1/2	1/4	1/8	1/16	*U. fasciata*
	169	84.5	42.25	21.12	10.56	(µg/ml)
A	+++	+++	+	-	-	
B	+++	+++	-	-	-	
AB	+++	+++	+/-	-	-	
O	+++	+++	+/-	-	-	
	1	1/2	1/4	1/8	1/16	*G. corticata*
	109	54.5	27.25	13.62	6.81	(µg/ml)
A	+++	+++	++	-	-	
B	+++	+++	++	-	-	
AB	+++	+++	++	-	-	
O	+++	++	-	-	-	
	1	1/2	1/4	1/8	1/16	*E. intestinalis*
	150	75	37.5	18.75	9.37	(µg/ml)
A	+	-	-	-	-	
B	-	-	-	-	-	
AB	+	-	-	-	-	
O	-	-	-	-	-	

Table 11.2: Hemagglutination titer

Blood type Rh+	*Protein extract (µg/ml)*		
	U. fasciata	*G. corticata*	*E. intestinalis*
A	42.25	27.25	150
B	84.50	27.25	NA
AB	84.50	27.25	150
O	84.50	54.50	NA

U. fasciata and *G. corticata* produced agglutination in Rh+ B as well as O erythrocytes.

DISCUSSION

Protein extracts from *U. fasciata* and *G. corticata* produced agglutination in all 4 human blood groups and thus showed no specifity towards any blood group within the tested concentrations. However, the extract from *E. intestinalis* showed partial specificity to the A and AB blood types. Human B erythrocytes have been reported to be relatively more sensitive to a number of algal lectins (Chiles and Bird, 1990). Our observations did not indicate any such specificity with the marine algae tested.

The minimum amount of protein required to produce agglutination was lowest for *G. corticata*. However, the same was relatively high as compared to the data that Dalton et al. (1995) obtained for other species of marine algae.

Faberglas et al., (1985) and Chiles and Bird (1989) have reported that rabbit erythrocytes may be the most suitable cell type in the initial screening for new agglutinins due to increased sensitivity as compared to human erythrocytes. No comparative study of animal blood cells was included in our study. Trypsinization of erythrocytes is also reported to enhance agglutination (Lis and Sharon, 1973) and inclusion of this step may result in a more sensitive assay suited to screening with low protein concentrations.

Preliminary observations from our results indicate inhibition of hemagglutination by all six monosaccharides tested. It may be premature to comment on any single sugar specificity of *U-fasciata* extract, as more sensitive tests using complex glycoproteins for inhibition must be conducted. These results are consistent with the findings of Hori et al. (1986). Rogers and Hori (1993) observed that algal lectins tend to show specificity with complex oligosaccharides within glycoproteins.

This study reports the presence of hemagglutinins in extracts from 3 algal samples. The competitive inhibition of hemagglutination by the monosaccharides indicates that the extracts are probably lectins or lectin-like substances.

REFERENCES

Bird, K.T., Chiles, T.C., Longley, R.E., Kendrick, A.F. and Kinkema, M.D. 1993. Agglutinins from marine macroalgae of the southeastern United States. J. Appl. Phycol., 5: 213-218.

Boyd, W.C., Almodovar, L.R. and Boyd, L.G. 1966. Agglutinins in marine algae for human erythrocytes. Transfusion, 6: 82-83.

Blunden, G., Rogers, D.J., and Farnam, W.F. 1975. Survey of British seaweeds for hemagglutinins. Lloyda 38 : 62.

Chiles, T.C. and Bird, K.T. 1990. *Gracilaria tikvahiae* agglutinin. Partial purification and preliminary characterization of its carbohydrate specificity. Carbohydr. Res. 297: 319-326.

Chiles, T.C., and Bird, K.T. 1989. A comparative study of animal erythrocyte agglutinins from marine algae. Comp. Biochem Physiol 94B : 107.

Colucci, G., Moore, J.G., Fieldman, M. and Chrispeels, M.J. 1999. CDNA cloning of FRIL, a lectin from *Dolichos lablab*, that preserves hematopoietic progenitors iTN suspension culture. Proc. Natl. Acad. Sci. USA. 96: 646-650.

Dalton, S.R., Longley, R.E. and Bird, K.T. 1995. Hemagglutinins and immunomitogens from marine algae, J. Mar. Biotechnol., 2: 149-155.

Etzler, M.E., 1986. In: The lectins: Properties, functions, and applications in biology and medicine, (eds). Liener, I.E., Sharon, N., Goldstein, I. Academic Press, New York: 371.

Fabergas, J., Llovo, J. and Munoz, A. 1985. Hemagglutinins in red sea weeds. Bot. Mar., 28 : 517-520.

Hori, K., Miyazawa, K. and Ito, K. 1986. Preliminary characterization of agglutinins from seven marine algal species. Bull. Jpn. Soc. Sci. Fish. 52: 323-331.

Hori, K.H., Matsuda, Miyazawa, H. and Ito, K. 1987. A mitogenic agglutinin from the red alge *Carpopeltis flabellate*. Phytochemistry, 26: 1335-1338.

Lis, H. and Sharon, N. 1973. The biochemistry of plant lectins (Phytohemagglutinins). Ann. Rev. Biochem., 42: 541-573

Rogers, D.J. and Hori, K. 1993. Marine algal lectins: New developments. Hydrobiologia 260/261 : 589 Quoted by Dalton et al. 1995.

Wright, A.E. and Douglas, S.R. 1903. An experimental investigation of the role of the body fluids in connection with phagocytosis. Proc. R. Soc. London B. 72: 357-370.

Chapter 12

Intracellular and Associated Marine Bacteria in the Sponge *Halichondria panicea:* A Potential for Pharmaceuticals

G.Gerdts, A.Wichels, H.Doepke and **C.Schuett***
Alfred-Wegener Institute for Polar and Marine Research
Biologische Anstalt Helgoland, 27498 – Helgoland, Germany
*Email: cschuett@awi.bremerhaven.de

ABSTRACT

Interactions between pro- and eukaryotes are diverse in the marine environment. Organisms of complex natural communities provide a broad scope of bioactive compounds. Interactions between pro- and eukaryotes of the North Sea and search for the bioactive compounds have been elucidated, using bacterial isolates derived from the sponge *Halichondria panicea*. Endocytic and associated bacteria have been detected microscopically and isolated after determination of a physico-chemical microenvironment by microelectrodes. Eukaryotic material containing both associated and endocytic bacteria has been subjected to amplification of its eubacterial and archaeal DNA by PCR. Data of denaturant gradient gel electrophoresis (DGGE) provide information on the community structure and bacterial diversity. Taxonomic determination by sequencing bacterial 16S rDNA reveals the presence of mostly unknown bacteria. Bacteria isolated from sponge *H. panicea* synthesize neuroactive compounds, which open up the calcium channel in cortical rat neurons.

Key words: Endocytic and associated bacteria, PCR amplification, DGGE-neuroactive compounds

INTRODUCTION

Nature provides a tremendous supply of different classes of chemical compounds, which are used as food or its additives, enzymes, components for the cosmetic industry, agrichemicals and pharmaceuticals. The list of pharmaceuticals is impressive. It includes antibiotics, cytostatic drugs, drugs against different (tropical) parasites such as virus, bacteria, protozoa, helminthes and so on. For centuries, bioactive components were extracted from plants and animals. That kind of extracts were used in traditional medicine in all parts of the world. The value of marine organisms was discovered only some decades ago. This investigation concentrated on invertebrates (Bryozoa, Cnidaria, Echninodermata, Mollusca) of the warmer regions (Reid et al., 1993; Unson and Faulkner, 1993; Schmidt et al., 1997; Pawlik et al., 2002; Salomon and Faulkner, 2002; De Jesus and Faulkner, 2003). Recently, it became clear that organisms from moderate marine environment also provide a broad spectrum of very potent bioactive chemical structures. The findings include the anti-tumor bryostatins (Pettit et al., 1991), extracted from bryozoa *Bugula neritina*, the anti-lung-cancer dolastatines (Bai et al., 2001), extracted from sea hare *Dolabella auricularia* or the broad spectrum didemnines (Look et al., 1986) for treatment of encephalomelytis, rift-valley and yellow fever, extracted from ascida of the genus *Didemnum*. All these organisms and more are common in the North Sea. Hence, most of the eukaryotic organisms (Groepler and Schuett, 2003; Kirchner et al., 1999, 2001) harbor specific associated and endocytic bacteria are also expected to occur. Sponges, for example, contain upto 50% of non-sponge biomass (Wilkinson, 1978). The guests are members of various groups of microorganisms, protozoa, phyto- and zooplankton as also small animals.

Until recently, the database was limited. The sponge *Theonella swinhoe* harbors bacteria which produce cytotoxic swinholides (Bewley et al., 1996). It is not clear whether the host invertebrates or their microbial guests produce the active compounds. Nevertheless, it can be assumed that associated and intracellularly living microorganisms play a key role in providing a wide scope of bioactive metabolites (Renner et al., 1999; Cheng et al., 1999; Engel et al., 2002; Tan et al., 2003). Particularly, bacteria from extreme environment are fascinating candidates in biotechnology. These organisms produce stable enzymes at high temperature of the deep sea vents, at extremely low temperature of arctic and antarctic regions or at high pressure in the deep sea.

This work focuses on the exciting facet of the interactions between the sponge *H. panicea* and its bacteria. The investigation describes methods for isolation and cultivation of microbes by simulating the natural conditions and the neuroactivity of bacterial isolates which are assigned to the CFB phylum *Cytophaga* group of proteobacteria. Data of the neuroactive compounds were already published (Provic et al., 1998).

MATERIAL AND METHODS

Habitats and organisms

The moderate marine environment of the North Sea is characterized by its high biodiversity. However, the productivity and the biomass of most of the organisms are low. Samples of *H. panicea* were collected by divers of our institute during several research cruises in the North Sea and the waters around the island of Helgoland.

Detection, isolation and culture media

Associated bacteria and intracellular bacteria of the sponge *H. panicea* were detected mainly by light microscopy.

To simulate natural environmental conditions, the physico-chemical parameters (i.e. O_2, H_2S, pH and redox potential) were determined inside and in the vicinity of the host animals (Revsbech and Jorgensen, 1986). Microelectrodes of ca. 15-50 µm diameter (Mascom Corp.) provided the first ecological information, which is crucial for the isolation of the microorganisms. Media were prepared from freshly blended and sterile filtered material of the sponges. Isolation and cultivation was carried out at semi-natural conditions by simulating the physicochemical conditions and the specific nutrient requirements of the microbes. Isolates were incubated at 16°C. Microaerophilic bacteria were cultivated in gas-controlled incubation chambers (MACS 500 glove boxes, DW Inc.). The gas components oxygen, nitrogen and CO_2 were added according to the data provided by the microelectrodes; these data represent in situ requirements.

BACTERIAL DIVERSITY AND IDENTIFICATION

Nucleic acid extraction

Extraction of nucleic acids was generally performed by a modified protocol of Anderson and McKay (1993), omitting the NaOH and HCl steps. Bacterial DNA was extracted either from bacterial biomass obtained from tissue material or from channel fluid of the sponge *H. panicea*. All DNA extracts, finally kept in TE-buffer, served as template DNA for PCR amplification. Prior to DNA amplification, extracts were analyzed by agarose gel electrophoresis on 0.8% agarose gels. From the intensity of the DNA bands, the required sample volumes for the PCR amplification were estimated.

PCR

To elucidate the microbial biodiversity of bacterial isolates derived from

Halichondria panicea, different sets of primers representing specific regions of 16S rDNA were used. Eubacteria were monitored by applying the primers p2 and p3 (Lane, 1991; Muyzer et al., 1993). Primer malf was originally designed for in situ hybridization of marine bacteria belonging to the α-subclass of proteobacteria (Gonzáles and Moran, 1997), but is used as a primer. For detection of archaebacteria, a nested PCR was performed using the primer set arch21f/985r (DeLong, 1992) and parch340/parch519 (Ovreas et al. 1947). Clamp extension for the primer sets were chosen according to Muyzer et al. (1993). For cloning, purified DNA extracts from sponge tissue material was used. Primers 63f and 1387r and PCR conditions are described by Marchesi et al. (1998).

DGGE

Bacterial diversity was investigated by DGGE profiling. DGGE was performed using a Protean II electrophoresis system (Biorad) according to Muyzer et al. (1993).

Cloning of mixed template DNA

The purified amplicons (QIAquick spin colums, QIAgen) were cloned into the pGEM T-Vectorsystem (Promega).

DNA-sequencing of PCR products and comparative sequence analysis

Purified plasmid DNA (Plasmid Isolation Kit, QIAgen) were sequenced according to the manufacturers' instruction on a Liqor DNA 4200 sequencer using the SequiTherm EXEL II long read sequencing Kit-LC (Biozym), as described by Seibold et al. (2001). For a rapid screening to distinguish unique and duplicate clones, sequencing was performed with a single dideoxynucleotide (ddG), as described by Schmidt et al. (1991).

Comparative sequence analysis

Sequences were aligned by the advanced BLAST search program of the National Center of Biotechnology Information (NCBI) website (http://www.ncbi.nlm.nih.gov) to find closely related sequences.

RESULTS AND CONCLUSIONS

Cultures

Marine bacteria are regarded as an important source for new bioactive components. Hence, biotechnology requires the "hardware" of bioactive microorganisms. However, the isolation of bacteria is crucial, since only 1-10% can be cultivated by traditional approaches. The highly specific ecological in situ conditions are mostly disregarded. Physico-chemical conditions (pH, O_2, H_2S and redox potential) required for cultivation of these bacteria are largely unknown. To optimize successful cultivation, natural conditions of the microenvironment have to be simulated. Microelectrodes are powerful tools which provide indispensable data for the microenvironment in different compartments of macroorganisms such as *H. panicea*.

Bacterial diversity and phylogenetic classification

Bacterial diversity in sponge *H. panicea* was determined by denaturant gradient gel electrophoresis (DGGE) profiling. This technique is based on the electrophoretic separation of DNA fragments containing different specific melting domains of high or low GC content. DGGE allows a direct and rapid overview describing the presence of different bacterial groups from natural environment. Fig. 12.1 displays the bacterial diversity of eubacteria, their subclass of α-proteobacteria and archaea in tissue material and canal fluid of sponge *H. panicea*. Section A shows the band patterns derived from samples of tissues (lanes 2, 4, 6, 8, 10) and channel fluid (lanes 3, 5, 9, 11). Each sample group demonstrates specific similarities of band patterns. Presumably, *H. panicea* harbors specific bacterial communities not only in tissue material but also in channel fluid. Section B shows the presence of the α-proteobacteria subclass. Tissue samples (lanes 2, 4, 6, 7, 8) display identical band patterns. This is also true for channel fluid samples (lanes 3, 5). From previous investigation, it is most probable that bacteria *Rhoseobacter* group, typical sponge organisms (Althoff et al., 1998) are present in these samples. Section C indicates the presence of archaea in tissue samples (lanes 2-6), which are nearly identical. Twenty bacterial strains isolated from sponge *H. panicea* were phylogeneticly classified by sequencing (Fig. 12.2). The dominant groups represent members of *Cytophaga/Flavobacteria* (7 isolates) and the subclass of γ-proteobacteria (7). Other groups were assigned to the α-subclass of (*Roseobacter sp.*) (3) and gram-positive *Actinobacteria* (2). Obviously the majority of the isolates are characterized by low similarities (<96%) to the type strains of data bank. It can be concluded that most of these isolates are new taxa. The screening of bacterial isolates for bioactive compounds is performed in cooperation with the Bayer Comp. Leverkusen, Germany; yet the promising data are confidential.

Nevertheless, bioactive compounds were detected earlier in two bacterial strains, isolated from sponge *H. panicea* in our laboratory. The investigation was performed in cooperation with the group of W.E.G. Mueller from University of Mainz (Procic et al., 1999; Perovic et al., 1998). Bacteria were tentatively assigned to the species *Antarcticum vesiculatum* and *Psychroserpens burtonensis*, both being members of the family *Theobacteriacea*. The isolates contained neuroactive compounds in their supernatants agonizing the *N*-

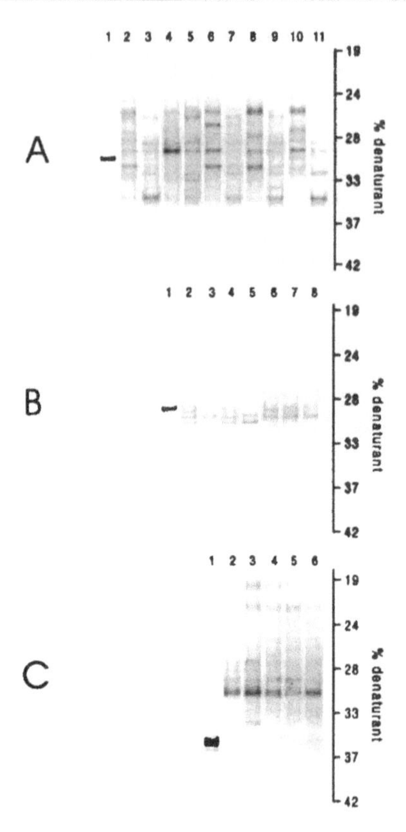

Fig. 12.1: Analysis of microbial diversity in the sponge *Halichondria panicea* (sampling site Helgoland) by DGGE profiling of 16S rDNA fragments obtained after the PCR amplification. (A) Eubacteria; specific primers P2 and P3; lane 1: control *E. coli* J53; lanes 2, 4, 6, 8 and 10: tissue samples; lanes 3, 5, 7, 9 and 11: channel fluid samples. (B) α-Subdivision of proteobacteria; specific primers MALF and P3; lane 1: control *Ruegeria algicola* DSMZ #10251, (German Type culture Collection, Braunschweig); lanes 2, 4, 6, 7 and 8: tissue samples; lanes 3 and 5, channel fluid samples. (C) Archaeal communities; specific primers arch21f / arch958r (outer primers) and parch340clamp / parch519r (nested PCR); lane 1: control *Halobacterium salinarium*, DSM # 97. Lanes 2 – 6: tissue samples (sampling date 22.07.98).

methyl-D-aspartate (NMDA) receptor. The response of this receptor is followed by a strong increase of intracellular [Ca^{2+}] in cortical neurons from rats as cell system. The receptor is involved in pathophysiological events of high scientific and medical importance like neuropathological changes such as AIDS, dementia, Alzheimer disease and prion diseases.

The isolation of intracellular bacteria is still a great challenge in microbiology. This requires new isolation strategies. The use of microelectrodes for the determination of the physico-chemical in situ conditions is a reliable option. Additionally, the natural nutrient composition should be simulated in cultures. The isolation of new species can be expected.

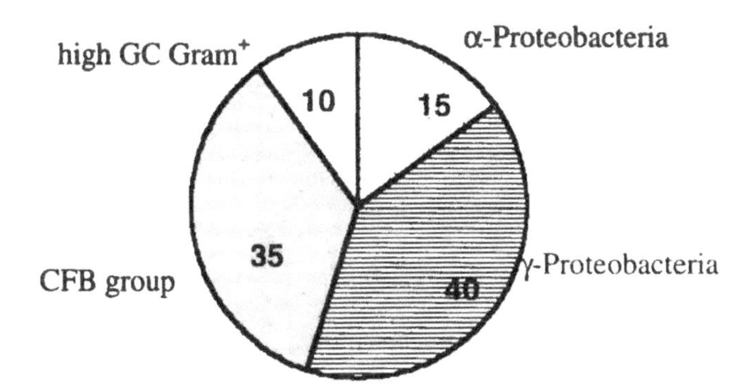

Fig. 12.2: Assignment of twenty bacterial isolates from the marine sponge *Halichondria panicea* (sampling site Helgoland), numbers in each segment represent the percentage of affiliation of the strains. (CFB : *Cytophaga/ Flavobacteria Bacteriodes* phyllum)

Marine associated and intracellular symbiotic or parasitic bacteria play a key role as bioactive source in biotechnology. Symbiosis research has a bright future. The elucidation of the manifold biochemical interaction mechanisms, the transfer of signal compounds involved and the advance of molecular techniques will provide new insight in the ecological strategies of symbiosis. This will enhance progress in basic research substantially. From the applied point of view, many new and helpful chemical compounds are awaited.

ACKNOWLEDGEMENTS

This work is supported by a grant from Federal Ministry of Education and Research (BMBF, Germany) and the Bayer AG Leverkusen (Germany).

REFERENCES

Althoff, K., Schuett, C. Krasko, A., Steffen, R., Batel, R., and Mueller, W. 1998. Evidence for symbiosis between bacteria of genus *Rhodobacter* and the marine sponge *Halichondria panicea* harbor also putatively-toxic bacteria. Mar. Biol. 130: 529-536.

Anderson, D.G and Mc Kay, L.L. 1993. Simple and rapid method for isolating large plasmid DNA from *Lactic streptococci*. Appl. Environ. Microbiol. 46: 549-552

Bai, R., Verdier-Pinard, P., Sanjeev, G., Stessman, C.C., McClure, K.J., Sausville, E.A., Pettit, G.R. Bates, R.B. and Hamel, E. 2001. Dolastatin 11, a marine depsipeptide, arrests cells at cytokinesis and induces hyperpolymerization of purified actin. Mol. Pharmacol. 59: 462-469.

Bewley, C.A. Holland, N.D. and Faulkener, D.J. 1996. Two classes of metabolites from *Theonella swinhoe* are localized in distinct populations of bacterial symbionts. Experientia, 52: 716-722.

Cheng, X.C. Jensen, P.R. Fenical, W. 1999. Luisols A and B, new aromatic tetraols produced by an estuarine marine bacterium of the genus *Streptomyces (Streptomycetales)*. J. Nat . Prod. 4: 608-610.

De Jesus, R.P. and Faulkner, D.J. 2003. Chlorinated acetylenes from the San Diego sponge *Haliclona lunisimilis*. J. Nat. Prod. 66: 671-674.

DeLong, E.F. 1992. Archaea in coastal marine chemicals. Proc. Natl. Acad. Sci. USA, 89: 5685-5689.

Engel, S., Jensen, P.R. and Fenical, W. 2002. Chemical ecology of marine microbial defense. J. Chem. Ecol. 28: 1871-1985.

Gonzáles, J.M. and Moran, M.A. 1997. Numerical dominance of a group of bacteria in the α-subclass of class Proteobacteria in coastal seawater. Appl. Environ. Microbiol., 63: 4237-4242.

Groepler, W., and Schuett, C. 2003. Bacterial community in the tunic matrix of a colonial ascidian *Diplosoma migrans*. Helgol. Mar. Res. 57: 139-143.

Kirchner, M., Sahling, G., Schütt, C., Dopke, H., and Uhlig, G. 1999. Intracellular bacteria in the red tide-forming heterotrephic dinoflagellate *Notctiluca scinillans*. Archiv Hydruilogie 54: 297.

Kirchner, M., Wichels, A., Seibold, A., Sahling, G., Schuett, C. 2001. New and potentially toxic isolates from *Noctiluca scintillans* (Dinoflagellata). Proc. on Harmful Algae, IX Int. Conf. on Harmful Algal Blooms, Tasmania 2000, 379-382.

Lane, D.J. 1991. 16S/23S rRNA sequencing. In: Stakebrandt E. and Goodfellow, M, (eds) *Nucleic acids and Techniques in Bacterial Systematics*. John Wiley, Chichester, pp. 115-175.

Look, S.A., Fenical, W.Q., Jacobs, R.S. and Clardy, J. 1986. The pseudopterosins: Anti-inflammatory and analgesic natural products from the sea whip *Pseudopterogorgia elisabethae*. Proc. Nat. Acad. Sci. U.S.A 83: 6238-6240.

Marchesi, J.R., Sato,T., Weightman, A.J., Martin, T.A., Fry, J.C., Hiom, S.J., and Wade, W.G. 1998. Design and evaluation of useful bacterium specific PCR-primers that amplify genes coding for bacterial 16S RNA. Appl. Environ. Microbiol. 64: 795-799.

Muyzer, G., De Waal, E.C. and Uitgerlinden, A.G. 1993. Profiling of complex microbial populations by denaturing gradient gel electrophoresis analysis of polymerase chain reaction-amplified genes coding for 16SRNA. Appl. Environ. Microbiol. 59: 695-700.

Pawlik, J.R., McFall, G. and Zea, S. 2002. Does the odor from sponges of the genus *Ircinia* protect them from fish predators? J. Chem. Ecol. 28:1103-1115.

Perovic, S., Wichels, A., Schuett, C., Gerdts, G., Pahler, G., Steffen, R. and Mueller, W.E.G. 1998. Neuromodulatory compounds produced by bacteria from the marine Sponge *Halicondria panicea* of the glutamate channel. Environ. Toxicol. Pharmacol. 6:125-133.

Pettit, G.R., Gao, F., Sengupta, D., Coll, J.C. Herald, C.L., Doubek, D.L., Schmidt, J.M., Van-Camp, J.R., Rudloe, J.J, and Nieman, R.A. 1991. Isolation and structure of bryostatins 14 and 1. Tetrahedron 47: 3601-3610.

Procic, I., Bruemmer, F., Brigge, T., Goertz, H.D., Gerdts, G., Schuett, C., Elbraechter, M., and Mueller, W.E.G. 1999. *Prorocentrum lima* (Dodge) associated with bacteria of the genus *Roseobacter*: Possible link to their toxicity. Protist 149: 347-357.

Ovreas, L., Forney, L., Daae, F.L. and Torsvik, V. 1997. Distribution of bacteriaplankton in meromictic lake saelenvannet, as determined by denaturing gradient gel electrophoresis of PCR-emplified gene fragments coding for 165rRNA. Appl. Environ. Microbiol. 63: 3367-3373.

Reid, R.T., Live, D.H., Faulkner, D.J., and Butler, A. 1993. A siderophore from a marine bacterium with an exceptional ferric ion affinity constant. Nature. 366: 455-458.

Renner, M.K., Shen, Y.-C., Cheng, X.C., Jensen, P.R., Frankmoelle, W., Kauffmann, C.A., Fenical, W., Lobkovsky, E., and Clardy, J. 1999. Cyclomarins A-C, new antiflammatory peptides produced by a marine bacteriaum (*Streptomyces sp.*). J. Am. Chem Soc. 121: 11273-11276.

Revsbech, N.P. and JØrgensen, B.B. 1986. Microelectrodes: Their use in microbial ecology. In: Advances in Microbial Ecology (ed.). Marshall. K. Plenum New York, 293-352.

Salomon, C.E. and Faulkner, D.J. 2002. Localization studies of bioactive cyclic peptides in the ascidian *Lissoclinum patella*. J. Nat. Prod. 65: 689-692.

Sambrook, J., Fritsch, E.F. and Maniatis, T. 1989. Molecular cloning: A laboratory manual (2nd ed). Cold Spring Harbor Laboratory Press, Cold Spring Harbor, New York Vol.I p. 646.

Schmidt, T.M. and DeLong, D.F. and Pace, N.R. 1991. Analysis of marine picoplankoton community by 16S rRNA gene cloning and sequencing. J. Bacteriol. 173: 4371-4378.

Schmidt, E.W., Harper, M.K. and Faulkner, D.J. 1997. Mozamides A and B, cyclic peptides from a theonellid sponge from Mozambique. J. Nat. Prod. 60: 779-782.

Seibold, A., Wichels, A. and Schuett, C. 2001. Diversity of endocytic bacteria in the dinoflagellate *Noctilucua scintillans*. Aquatic Microbiol. Ecol. 25: 229-235

Tan, L.T., Cheng, X.C., Jensen, P.R. and Fenical, W. 2003. Scytalidamides A and B, new cytotoxic cyclic heptapeptides from a marine derived fungus of the genus *Scytalidium*. J. Org. Chem. 68: 8767-8773.

Unson, M.D. and Faulkner, D.J 1993. Cyanobacterial symbiont biosynthesis of chlorinated metabolites from *Dysidea herbacea*. Experientia. 4: 349-353.

Wilkinson, C.R. 1978: Microbial Associations in Sponges. 1. Ecology, physiology and microbial populations of coral reef sponges. Mar. Biol. 49: 161-167.

Subject Index

Androgenesis 39
AFLP markers 33
Antibody resistant strains 124
Antigen encoding gene 134,135
Anti-freeze protein gene 6
Apoptosis 74
Aqua-advantage TM 4
Aromatase 70
Asymptomatic disease outbreak 126

β-actin promoter 6
Bi-cistronic vector 8
Bioactive compounds 155
Bootstrap values 35
Biodegradable microspheres 85
Brain-pituitary-gonadal web 83
Branchial hyperplasia 129
Broad spectrum dideamines 152

Cadaveric sperm 46
Coagulable negative organisms 127
CHH family gene 109
Cloned GnRH-R 92
Co-valent cross linking 22
Cryopreservation 45

Denatured gradient gel electrophoresis 155
Di-spermic activation 40,50
DNA cpG motiffs 135
DNA nuclear import 25
DNA vaccines 133
Double stranded RNA 111

Electroporation 8,139
Endogenous GnRHs 87, 93
Environmental sex differentiation 66
ER α-immunoreactive cells 73
Estrogen receptor 70
Estrogen responsive element 87
Eyestalk ablation 106

Fluorescence in-situ hybridization 24

Gene bombardment 10
Gene gun delivery 23
Genomic impurity 41
Genomic integration 3
Genetronics 139
Global warming 76
Glycoprotein gene 136
Gonadotropin (GtHs) 83, 89
Gonadotropin cells 7, 72
GnRH 83
GnRH agonist (GnRHa) 85,92
GnRHa-Evac implant 86
GnRH triggered signal mechanism 92
GnRH receptors (GnRH-R) 83, 90, 91
Growth acceleration 13
Growth inhibiting hormone 106, 108
GTH cells 71

Halichondria panicia 151
Hemagglutination assay 147
Heterospecific insemination 47, 49
Homeopathic induction 119
Hormone receptor 89
Hyper glycemic hormone (CHH) 107

Immersion vaccination 139
Immunocytochemistry 71
Immuno compromise 128
Indigenous GnRH genes 87
Immunoreactive gonadal cells 71,72,73
Internal ribosomal entry sites 1
Interspecific androgenesis 58
Interspecific cloning 38, 59
Intracellular bacteria 156

Juvenile hermaphroditization 74

Karyophilic proteins 22

Lectins 146
Luciferase expression 26

Maternal genome inactivation 44
MHC analysis 39
Microbial diversity 155
Micro electrodes 155
Microinjection 8, 9, 22
Microspheres 139
Molecular taxonomy 31
Mosaicism 3
Moult-inhibiting hormone 106, 108
Mycological examination 128

Natrum muriaticum 120
Nei's genetic distance 34
Nuclear localization 22
Nuclear receptors 115
Neuropeptides 87, 89
Neurotransmitters 87, 89

Particle bombardment 138
Particle-mediated gene transfer 26
Passive immunization trials 110
Photo-reactivation 43
Polyspermy 48

RAPD fingerprints 31
Receptor modified endocytosis 136
Reporter gene 7,134, 138
Ribonucleic protein assay (RPA) 88
Rock lobster 112

Sex control 66
Sex differentiation 66,68, 74
Sex-specific GnRH cDNA 88
Sex-specific marker 55
Sinuos gland peptides 108
Signal peptides 21, 88
Spawning technology 84
Species-specific marker 52
Stage specific embryonic mortality 57
Supermales 53

Transgenesis 1, 24
Tc1 transposon specific primers 53
T-lymphocytic response 133

Vectors 133
Vibrio harveyii biotypes 134, 135
Vibriosis 135

UPGMA dendrogram 31
UV-irradiation 43

For Product Safety Concerns and Information please contact our EU representative GPSR@taylorandfrancis.com Taylor & Francis Verlag GmbH, Kaufingerstraße 24, 80331 München, Germany

T - #0061 - 160425 - C12 - 244/155/10 [12] - CB - 9781578083725 - Gloss Lamination